Let's Fix This

Cleaner Living in a Dirty World

Chandra Clarke

OBSIDIAN OWL
PRESS

Special discounts are available on quantity purchases by corporations, associations, and others. Contact customerservice@tigermaplepublishing.com for details.

Paperback ISBN: 978-1-7772174-9-5

For my children. And for yours.

Contents

Introduction

> *'I wish it need not have happened in my time,' said Frodo. 'So do I,' said Gandalf, 'and so do all who live to see such times. But that is not for them to decide. All we have to decide is what to do with the time that is given us.'*
>
> — *JRR Tolkien, Lord of the Rings*

The headlines are full of scary images and statistics. Wildfires. Hurricanes. Collapsing ice shelves. Uprisings and protests. Creeping autocracy. All made worse by the surreality of the lingering echoes of the pandemic.

I don't know about you, but I haven't been sleeping well. Or, you know, at all.

Our governments seem to sit on their collective hands and do nothing, or worse, actively roll back progress. Social media has become a doomscroll; a toxic soup of negativity, conspiracy theories, and snark. The future seems uncertain.

The problems seem so BIG. It's hard to know where to even start. What can I do? Is there any hope?

In my heart of hearts, I believe there is.

Here's why: When I started looking for the glimmers of light in the darkness, looking beyond the headlines, I discovered that there *are* people trying to fix this. Hundreds of thousands of people all around the world, just like you and me, who have rolled up their sleeves and are *doing* something. Taking action in their homes. In their places of work. In their communities.

There are solutions. We *can* fix this. History has shown us that when we listen to the best scientific advice of the day and do something about a problem, we can halt the damage and even reverse it. Climate change deniers like to point to acid rain and the ozone 'hole' as things we got alarmed about that didn't happen, but in fact, they were enormous problems that we figured out and we *fixed*.

But you can't do it alone. None of us can. It's time for all of us to link up and start pushing hard in the same direction.

This isn't about activism anymore. Or 'raising awareness.' It's about *action*.

We've got to get organized.

Let's find out how, together.

What will follow this introduction is a series of brief chapters about practical actions you can take *right now*, today, tomorrow, and the next day. Some of them will be easy, some of them will be harder, some will be free, and some of them will take money. But they will all be things *you* can do.

We'll take it one step at a time.

How to use this book

This book covers a lot of ground.

We have a lot to do! It feels overwhelming.

But that's why I have broken things out into a) systemic things you can help change and b) individual changes you can make, all in bite-sized steps.

Indeed, you might feel at first that some steps are just a little too small. In the individual changes portion (Part II), some of what I break out might seem obvious in retrospect. And it is, *when you think about it*. But one problem we face is that our lives are so incredibly busy. One of our coping methods is to pick the fastest, cheapest, most marketed option available to us. This book is designed to get you to see things fresh, in the comfort of your home, and give you the time to think and plan.

You can't fix everything overnight, but you can do something right now. And then the next thing, and the next thing, and before you know it, you'll have made a major difference.

So, I would suggest you read the introduction, this section, and the Part I and Part II explainers to start with. Then you can either pick a chapter heading that appeals the most and start with that. And of course, you can always just start working your way through the book methodically.

But here's the really important part: set a calendar reminder to come back to the book.

Why? Because again, our lives are busy, and the entire system has evolved to distract you.

Remember, making your home and life more sustainable is a process. It won't happen overnight. But you need to start the journey.

Let's go!

Part 1
At The Systems Level

One's got to change the system, or one changes nothing.

— *George Orwell*

Recently, *The Guardian*[1] reported on a study that suggested that just 100 companies were responsible for 71 percent of greenhouse gas emissions.

100 companies. 71 percent.

Which sounds like an actionable bit of knowledge. Just boycott those 100 and we're good to go, right?

Except that it turns out those companies are coal, oil, and gas producers that you and I don't buy from *directly*.

So, that means we must take a two-pronged approach:

We need to make systemic changes.

We need to make personal changes.

Systemic changes are up first, because while we can definitely influence things by making small personal changes, we must kick the *system* hard to get the actions we want as fast as we need them.

Let me use a real-world example to illustrate what I mean.

You probably already recycle your plastic bottles, right? Or maybe you don't even use them. That's awesome, and you should be congratulated. It's important that you keep refusing to use products like that.

But ... as long as the *system* continues to produce plastic bottles and doesn't account for the hundreds of thousands of people who either can't or won't recycle them, and also doesn't account for the bad actor companies that just dump what they're supposed to recycle, we still have a huge problem.

Well, I hear you say, we just need to get more people to recycle. Education! Awareness!

To which I say: hogsnarfle.

For one thing, we've been running education and awareness programs for *decades*. They only work if people are receptive to the message *and* can act on it.

But more to the point, there's just too many of us!!

Let's do some math. Don't worry, I'll do all the heavy lifting here.

Coca-Cola,[2] as recently as 2023, said that they sell 1.9 billion servings of their drinks in 200 countries every day.

Now, let's assume we do such a great job of recycling that a whopping 90% of those serving containers are recycled.

That still leaves one hundred and ninety *million* containers in our landfills, oceans, and by the side of the road.

190,000,000.

Every. Single. Day.

(And yes, you could argue that not all those servings are individually packaged, what with bigger bottles and fountain drinks, but even if you slash that number in half or by two-thirds, it's still mind-bogglingly huge. It's also just the drinks from one company. Think how many other products there are out there?!)

Now let's be honest with ourselves here. You've seen the comments on any social media post. You know what your co-workers and the people in your community can be like.

Can you really see any future in which we have 90% of people doing the right thing 100% of the time?

No, me neither. And I'm an optimist!

So that's why we need to tackle the systems first. We

must have systems that account for *real human behaviours*, both good and bad.

Some Important Systems Level Concepts

> *We live in capitalism, its power seems inescapable — but then, so did the divine right of kings. Any human power can be resisted and changed by human beings. Resistance and change often begin in art. Very often in our art, the art of words.*
>
> *— Ursula K. Le Guin*

If you've taken part in any online discussions about the environment, capitalism vs. socialism, or any related topic, inevitably someone will post one or more of the following comments:

Eat the rich!

Abolish the dollar!

Burn it all down!

These glib comments feel revolutionary in the moment, but don't actually move the needle, because they don't answer an important question:

And then what?

Sure, we can destroy the establishment, but what do we replace it with? Before you answer with your favourite political system: what do we replace it with that doesn't immediately (or shortly thereafter) revert to the same system with a different name?

One reason our current problems seem so hard to solve is that most of us cannot visualize something different. What does better look like?

Some people are doing the high-level thinking on these questions, and I want to introduce you to some concepts that I think are relevant here. A full treatment of each is beyond the scope of this book; my goal here is to get you to shift your mindset and start thinking about how we can do things differently. I want to get you excited about the possibilities, rather than fearful about the uncertainty.

Exploitative vs. regenerative or creative

Historically, human development and industrial progress have used an exploitative model. Materials are extracted, used, and then discarded. The implicit assumptions are infinite resources and exponential growth.

Regenerative systems try to offer a holistic and interconnected approach to living and working with the natural world. Regenerative practices go beyond plain old sustainability. They aim not just to reduce harm but to actively improve, restore, and revitalize ecological health.

This involves creating systems that mimic nature's processes, where waste for one is a resource for another, and where activities contribute positively to the environment.

The shift toward regenerative systems requires a fundamental change in how we view our relationship with nature. Instead of seeing the natural world as a set of resources to be exploited, we recognize it as a complex, living network we are a part of and dependent upon.

A simple example here would be a winery[3] that also keeps... ducks! The ducks eat the grass between the vines, keeping it short, and eat the bugs that would consume the grapes. They fertilize the soil with their poop and provide duck feathers and eggs. The need for pesticides and synthetic fertilizer is reduced or eliminated.

Related: circular vs. linear economies

A linear economy takes, makes, and disposes. It's a straight line from the mine shaft or oil well to the dump.

A circular economy starts at the drawing board: designing products and services that are *meant* to be recycled, repurposed, or reused. An example would be a modular cell phone, where the parts can be removed and swapped out for repairs and upgrades, and even put into other devices. So instead of getting rid of your phone every time there's new functionality, you'd simply upgrade the bits you wanted to upgrade, and the old parts would go to other people, or they'd be easy to break down into their components and recycled.

Another example, already happening[4], would be a deposit return and recycle system for drink and food packaging.

No more straight lines: everything is a closed loop.

Rewilding

Another important idea to cover is rewilding. This is a step beyond standard wildlife conservation methods, and the idea here is to return large spaces to their natural states as much as possible. While some keystone species are deliberately brought back to the space, after that, nature is allowed to get on with it, with less human management or interference.

Also, crucially, these areas are no longer available for human use of the land for either habitation or resource extraction. In some versions of the concept, the thinking is that we end up with human enclaves in much larger natural tracts than we have now. Less "dominion over the land" and more "sharing the space" with all the other plants and critters.

An example of rewilding would be the project that brought wolves back to Yellowstone Park[5] in the US, and then let the wolves get on with what they do, which is to catch and eat ruminant animals like deer. This has reduced overgrazing, restoring natural habitats for other species in Yellowstone. The entire area is now much healthier and biodiverse.

I've glossed over a lot of nuances here. There are dozens of variations of these concepts, they are not without their critics and concerns. This is also not a comprehensive catalog of some attempts to rethink our current systems.

The point is not to debate the merits of the concepts, but to broaden your horizons. I want you to see that things *could* be radically different from what they are now, and that there are people trying to come up with visions of what that looks like.

Local Government

> *Never doubt that a small group of thoughtful, committed citizens can change the world; indeed, it's the only thing that ever has.*
>
> — *Margaret Mead*

In the last section, we talked about how if we want to effect *genuine* change, we need to push for system-level change.

We're going to start with local government.

Why? Two reasons:

Your local, municipal-level government is the one that is going to be (or already is being) hardest hit by environmental problems, and the related social problems. Polluted water? A problem for your local water treatment plant. Flooding? Your local EMS teams. Heat waves? Also your local EMS teams. Terrible air quality? Your local hospitals, overcrowded and struggling. Too much trash? Your tax bill, inflated by waste management fees.

Your local government is also the one you have the most direct access to, especially if you live in a smaller community. So, how can we get systemic change at the local level?

First, pick a single issue and a desired outcome. Yes, there are a lot of things you could try to fix, but if you approach your local government with a laundry list, you'll get exactly nowhere. One thing at a time.

Here are some suggested issues and desired outcomes.

Plastic pollution: Ban the use of a plastic item (or even all single-use plastic items in your municipality). Sometimes people laugh at plastic straw or plastic bag bans as being silly and ineffective in the face of so much plastic everywhere else, but every time one of these goes into effect, it shrinks the market for plastic products and sends a message that we're fed up with plastic pollution. (Note: It's important to acknowledge that there are some people who need straws because of physical or mental constraints. Fortunately, there are now washable, reusable stainless steel and silicone straws available as replacements. I bring this to your attention to a) remind you that not everyone is able-bodied or enjoys the same circumstances you do and b) point out that if we put our minds to it, we can come up with good solutions.)

Air pollution: 'No car' zones in your municipality. This reduces air pollution, increases the use of public transportation, and starts forcing urban planners to make your city more walkable or better for bikes. (This also improves health outcomes[1], which means lower health care costs.).

Electrified transportation infrastructure: Are there enough public car chargers in your city? Are there bylaws in place to prevent fossil cars from parking in charging spots? How about bylaws to help condo and apartment dwellers secure charging spots? Are your public transit options based on renewable energy? Do you even have public transit options?

Electricity supply: Has your local utility upgraded to renewable energy or is it still burning coal? Are you allowed to set up your own microgrid or is that against the law?

Trash: Does your municipality have an organic waste (composting) system? Is recycling collected more often than trash? How much stuff can really be recycled in your city?

Native vs. invasive plants: Does your city have any bylaws about what plants the local nurseries can sell? If they're not selling native plants, can you at least get a ban on harmful, invasive plant sales?

Invasive plants part two: What steps is your city taking to tackle invasives like phragmites, dog strangling vine, knotweed (or whatever is a problem in your region)?

Plants, part three: Does your municipality have a tree replacement program? Cities are very good at chopping things down, what do they do when a problem tree is removed?

How to Take Action

There are three ways you can act at the local government level.

1. Email your representatives. Email them all, either all at once, or one per week. (Put it in your calendar!) Describe the problem, describe your solution, *and* show how your solution would benefit the taxpayer. Forward news stories from other cities where a solution has been implemented. If you can work out which bureaucrats and front-line workers would implement your solution, email them too.
2. Form an action group! Get a *bunch* of people emailing councillors, attending meetings, circulating a petition, chanting in front of city hall, and so on. Remember, there's strength in numbers! (Pro tip for getting other people on board: Invite them personally, don't just put out a general call for volunteers.)
3. Run for office. This might sound daunting, but you just need to remember that there's nothing special about the people who have run for office before. You don't even have to win a seat (although obviously it's better if you do). Running for council gives you the chance to go door-to-

door to talk to people about your issue and get them on board.

Remember: be specific. Don't just talk in general terms about zero waste or climate change. Give your local council something they can *action*. A proposal they can vote on. Give them something that would make them look good to voters.

One other important point: Government officials at any level often claim that they can't afford green initiatives. Help them do the math. Point out where investing in one area will help them reduce their expenses in other areas, or boost tourism, or bring in more tax revenue. We don't have to like the fact that we're ruled by economics, but we can use it to our advantage.

Regional Government

We cannot become what we want by remaining what we are.

— Max De Pree

In our last section, we talked about the need to push for change at the municipal level. Now we'll kick it up a notch and focus on regional government.

Depending on where you live, that could be your state, your province, your county, your canton, prefecture, etc. For the sake of simplicity, let's consider this any level of government between municipal and national.

These governments aren't as monolithic and hard to budge as the national governments can be, yet they quite often have big budgets and mandates to deal with environmental issues.

What should you press for at the regional level? Here are some suggestions:

- The preservation of, the expansion of, and/or the creation of conservation areas and parks.
- Subsidies for the development and installation of green economy infrastructure. This could include incentives to buy cars that aren't powered by internal combustion engines, setting up charger networks, or providing incentives to homeowners for conservation and power generation.
- How is power generated in your region? Is it green or something like coal? What are the regulations regarding 'going off the grid?' Is it possible to sell power generated at home? What can be done to empower (pun intended!) people in this area?
- How do your waterways seem? Are they clean? Which authority handles them? Is there a plan to fix problems like phosphorus runoff from farms, or pollution from factories and municipalities? How often are they inspected and by whom? Who's tapping into the water supply and are they paying a fair price to do so? (Looking at you, bottled water companies!)
- How well do your public transportation networks link up? Is it possible to travel everywhere in your region through public transit? If not, could it be possible? (The fancy name for this is intermodality, by the way.) What would it take to

make transitions from one public transit system to another seamless for the rider?

As with municipal-level issues, pick *one* issue to advocate for. You're more likely to run into issues with multiple governing bodies here, so take the time to research who handles what before you make noise.

How to Take Action

There are many ways you can act at the regional government level.

Find out who *else* might be making noise about your preferred issue already. Chances are, given that you're looking at the regional level, that there's some sort of advocacy group or non-governmental organization (NGO) already involved. Then join it and commit some regularly scheduled volunteer hours to making it succeed.

It might be the case that there isn't an organization already, or no local chapter. Guess who could start one? You!

You can make your views known to your regional representative. A phone call might be more effective here, as regional reps get a *lot* of email. You could even go old school and send a paper letter by post.

Don't forget running for office here too. Obviously, this is going to take more money and effort than it would at the municipal level, so it might not be an option for you.

But you could also consider joining a political party. Candidates and elected officials don't come up with policy on their own; they're just the public face of the party in their area. You can be one of the 'behind-the-scenes' people, pushing for policies at the party level between elections, keeping party members organized, helping to select candidates or delegates within the party that align with your issues, and helping to get your preferred candidate elected when the time comes. We talk a lot about how much money goes into political campaigns, but tens of thousands of hours of volunteer labour is also required. What could you contribute?

As with municipal level governments, regional reps will often make noise about budgets. Show them how they can't afford *not* to fix a problem[1]. Where I live, for example, I can demonstrate that investment in electric cars will reduce air pollution, which will reduce the provincial healthcare budget, and improve economic productivity. Why? Because air pollution makes many people sick (think of all the chronic respiratory conditions there are, from asthma to COPD) and I can find statistics online to show how much these conditions cost the jurisdiction.

National Government

The idea is to change the nature and value system of the nation as a whole.

— Sunday Adelaja

It's a frustrating time to be a progressive voter. On one hand, the existing power structures seem inclined to corporatism and incrementalism, when what we need are sweeping changes, fast. On the other hand, opposition parties in many countries right now seem *regressive*, and so the choice in too many nations seems to be to vote for the status quo and hope to inch it forward or risk sliding backward.

It doesn't help when world leaders spend a lot of time pointing fingers at each other, insisting that some other country needs to go first in cutting emissions and reducing pollution. While it's true that certain countries have outsized numbers because of the size of their population (e.g., China) or the size of their economy (e.g., the USA), the truth is *all* of us, particularly the rich, so called "developed" nations have terrible numbers on a per-person basis.

For example, Canada, at the time of writing[1], emits 14.2 tons per person versus India at 2.0 per person, and China at 8.0 per person. In fact, on this basis, Canada is worse than China and India combined! So don't assume your country is doing well based on overall emissions numbers. Dig into the data and use these numbers when you act. Speaking of which:

How to Take Action

1. Vote! The nail-biting closeness of the 2020 US elections is evidence enough that every vote counts. Make sure you vote in every election that you can, at every level, but especially the national level.
2. Work against gerrymandering, which can happen anywhere. Make sure your voting district boundaries haven't been tinkered with such that it favours one party over another.
3. If you can't vote to move forward, at least vote to prevent harm (to yourself and especially to vulnerable and marginalized populations). Especially consider what splitting the progressive vote might do in your district before marking that ballot.

4. Email or call your current national representative frequently and be sure they know how you feel about environmental issues. Sign those petitions, too.
5. Change things from the inside: join the party that has the best chance of forming a government and do the grassroots work to make the party platform progressive. For those of you who just scoffed, I get it. Burning down the existing system sounds much more appealing. But revolutions need a solid plan for what happens when the smoke clears; they usually don't and then the organization implodes, or the same old autocrats move back into the power vacuum. Alex Steffen has a good line here: "Cynicism is obedience."
6. Join big organizations that change things from the outside. Groups like the World Wildlife Federation, Greenpeace, and others have enough members (and funding) that they can browbeat governments into being better from time to time.
7. Protest! Join or organize *peaceful* demonstrations.

In our next section, we'll discuss the power behind the throne: big corporations.

Corporations

As long as politics is the shadow of big business, the attenuation of the shadow will not change the substance.

— *John Dewey*

Whether you're a dyed-in-the-wool capitalist or an ardent socialist, it's fair to say that large businesses — multinational corporations especially — have incredible power. It's also fair to say that power isn't always used for the greater good. This doesn't mean that businesses are inherently evil; indeed, there are Certified B corporations and social enterprises all over the world trying to do better. It's just that far too many companies are optimized for profit over people and the planet. And that results in bad outcomes.

John Bunzl and Nick Duffell, writing in *The Simpol Solution*, suggest that we are in the grip of what they call "destructive global competition" or DGC. As long as there is *one* jurisdiction that 'competes' by offering low corporate tax rates, lax environmental regulations, soft labour laws, or some other incentive for corporations, then all the other jurisdictions must do something similar. Otherwise, corporations move jobs elsewhere. No jobs result in no votes for politicians, and of course, unemployment for the citizen. It's a downward spiral.

DGC is why even progressive-leaning political parties seem to promote the status quo. They can only push so hard in this environment, before losing votes and thus losing the power to change things.

This must change. It is unsustainable, no matter what timeline we consider.

So, what can we do? As a consumer, we're often told we can punish companies by boycotting their products. That is still true in certain industries, especially for small- to medium-sized businesses, *if* you can get enough people on board for a boycott. I'm sure you can think of many examples where public outcry and a drop in sales reversed an announced policy or changed a specific company practice.

It's not effective in the long term, however, as public attention is fickle. Neither is it much of a bother for large or multinational corporations. Consider The Procter & Gamble Company, which owns something like 65 different brands, with dozens of individual products under each brand line. You could boycott one brand of product in favour of another, only to discover, once you read the label carefully, that they're both owned by P&G. So not only is it hard to avoid that company's products in the first place, but they have so many that a temporary drop in sales for one brand won't hurt them much.

And let's be real: boycotts are also only truly available to the privileged, those who have the time and the money to make considered purchases. A single mother working two part-time jobs, and who is reliant on public transit to get groceries, doesn't have as many choices as a double-income family with two personal vehicles. So, while you can avoid one corporation's products completely should you want to, that corporation can still *make millions of dollars* from other consumers. Thus, we need to do more than just temporarily avoid a product for a while.

How to Take Action

1. Complain *directly* to that company's customer service channel. A drop in sales can be attributed to too many other things: a dud advertising campaign, a new competitor, a sudden price sensitivity, the end of a trend. If you're going to stop buying a company's product, you must let them know why. (Note: Don't be rude to the customer service rep for any reason. They're just doing their job.) In a blog post, Canadian grocery store chain owner Galen Weston once noted that the angriest emails they'd ever received was when they started wrapping their cucumbers in plastic. A flood of

complaints *does* get attention, and sometimes even action.

2. Publicly call the company out on social media. Not only is it harder for a company to ignore something being done openly, but other people will also see you taking a company to task and might be encouraged to avoid the product and voice their displeasure as well.
3. Do both of the above even if you have never bought the product and don't intend to. For example, I never buy soda, but I have complained to soda manufacturers about their continued use of plastic O-rings to distribute their products. (Note: It's beyond frustrating that these things are still in use. We knew back in the 1980s that they actively harmed wildlife.) The airport phrase works here too: If you see something, say something.
4. Bring the problem to the attention of a larger organization or media outlet, too. For example, it wasn't widely known that retail clothing stores were destroying and dumping millions of unsold items, until it came to the attention of the media and advocacy groups. Now there's considerable pressure against these "fast fashion" practices.
5. Rethink your investments. If you are fortunate enough to have a retirement fund, have a hard look at where you're invested. There's no point in avoiding a company's products if you're actively supporting their current stock market valuation with your mutual funds! Many brokerages and

banks now offer "ethical portfolios," and while there's a lot of greenwashing out there, these are worth a look. Talk to your advisor about establishing an investment policy.

6. Push an investment entity to divest. When big investors, like the Vatican, for example, announce they're no longer going to invest in anything related to fossil fuels, you can bet it makes a difference. Pick a bank, a mutual fund, a private equity firm, your government's pension plan, and tell them to divest themselves of environmentally destructive companies.
7. Do the same to your insurance company. Big investments come with enormous risks, which big insurance companies underwrite. There have been several successful campaigns to pressure insurance firms into declining to insure things like oil pipelines.
8. Support eco and social justice lawsuits. Individuals don't have the resources to fight corporations in court, but non-governmental organizations (NGOs) do, especially if you donate to their legal funds.
9. Finally, rather than boycotting, which is temporarily abstaining from purchase, permanently and proactively switch to eco-friendly products and services. Indeed, in Part II we'll talk about how to do just that.

On the Job

> *The only thing necessary for the triumph of evil is that good men do nothing.*
>
> *— Edmund Burke*

In previous sections, we've talked about how we desperately need systemic changes, and in particular, changes in the way corporations do business.

And it's absolutely correct to focus on the damage large companies have done to our planet. For instance, you only have to look at the damage caused by the oil company Pemex, which set the freakin' *ocean* on fire[1], to see that there's a problem.

But it's not just corporations — small- to medium-sized businesses have environmental impacts too. And what are all businesses, but organizations made up of people? People like you and me?

Sure, you might not be a CEO, or in the C-Suite at all, or even at the middle-management level. But even if you're just a cog in the proverbial machine, you have a voice. And it's time we all started using our voices at work. As the folks at Project Drawdown[2] contend, every job must be a climate job.

Here are some ideas for using yours. Remember to frame your suggestions or push for changes in a way that demonstrates the benefits to the organization:

- Sit down with a pencil and paper (or your phone, or whatever) and start thinking objectively about how you do your job. Are there ways to make it cleaner and greener? (For example, if you are a courier, are you shutting down your vehicle when you make a delivery or are you leaving it to idle? What would happen if turning it off was company policy? How much pollution would that remove from the equation? How much would it save the company in fuel costs?)
- Think too, about your section, unit, division, department, or whatever they're called in your company. Maybe your specific job is pretty clean and green, but what about the bigger picture?
- What consumables is your organization constantly having to reorder? Could they be replaced with reusables? Or done away with?
- What happens to stuff that doesn't get sold at your retail job? Does it get destroyed? Is it thrown

in the dumpster? Can you push to have the stuff donated? Can you tally up how much the company is wasting and send a report up the chain?

- Are you in charge of ordering supplies? Can you influence what gets purchased? Can you push for more eco-friendly versions of the things you use?
- What processes routinely produce a lot of waste in your company? Can you fix the process, so it generates less waste in the first place? Can you reuse or recycle that waste?
- What's energy use like where you work? Are people constantly leaving monitors and lights on? Leaving windows open while the heat is on? What solutions can you propose at the next staff meeting that would be more efficient and save the company money? Posting passive aggressive signs around the office doesn't work, but maybe having, for instance, the IT department implement a computer network power saving policy that shuts things down automatically might.

So, we've covered some basic actions we can take to improve things. These are generally safe to do and might even be good for your career if you've done it in such a way as to show initiative and a willingness to help the company's bottom line.

What About Bad Companies?

But what if you're working for a company that is genuinely a bad actor rather than being merely careless? You know the kind: flouting regulations, dumping illegally, not looking after the health and safety of its workers or its community. Or even just in an industry that's been proven to be bad for people or the planet, even if they're theoretically doing everything "by the book?" (Remember that adage: Just because it's legal, doesn't mean it's right.)

What you do in these cases depends on your financial situation, your conscience, what legal protections you have and so on.

If it's safe for you to do so — that is, you believe you're able to push for change without losing your job and imperilling your family in the process, you could make those suggestions. (Just remember that businesses might demand loyalty, but rarely give it back, so assume you're not truly secure).

If it's not safe for you to do so, then your priority should be to *leave the job*. Yes, leave it. You don't have to be part of the problem. (And if you're a young person just joining the workplace, don't sign up to work for bad actors or industries in the first place.)

That might mean sacrificing evenings to attend night school and reskill; or if you already have transferable skills, start applying for new jobs; and getting your finances in order so you won't be bankrupted by a change in, or complete loss of salary. (This is just a good idea anyway, as we never know when a company might dump us.)

In other words, find a way to get off the treadmill that forces you to accept a status quo that is endangering us all. Pulling in big bucks and being able to afford that house or nicer car won't help you when your town sets temperature records and then burns to the ground[3], or when there are food shortages because you have a record-setting plague of mice.[4]

Now I'll take a minute here to acknowledge that some of you reading this genuinely might not have a choice about where you work, for a variety of reasons. Maybe a family caregiver situation has you pinned down, or perhaps you live in a small town and the only major employer is a bad actor. It could very well be that you're scraping by and getting the time and money to re-skill isn't happening. I get it. I see you.

Whatever the reason, do what you can to lay the groundwork to freedom, if not for yourself, but so that your children have better options. And otherwise do what you can do as a volunteer in other areas.

And while I would never suggest that anyone break the law, I will say that whistleblowers are key to getting corporations and systems to reform. Fortunately, for anyone contemplating blowing a whistle on the activities of their organization, there are lots of articles available online on how to do so and protect yourself. Make sure you search for how to do it in your country, as protections (if they exist at all) vary from nation to nation and sometimes region to region.

Don't forget that there are search engines like DuckDuckGo that don't track your activities, or special browsers like Tor, which helps you stay anonymous online. And there are a lot of news organizations that have anonymous tip boxes available too. Regardless of the tools you use, never do this at your place of employment or with company equipment, even at home. Assume the IT department is monitoring your Internet and computer resource usage.

Donations and Investments

> *I have found that, among its other benefits, giving liberates the soul of the giver.*
>
> — *Maya Angelou*

Giving

Understanding the role of donations and investments in combating climate change involves recognizing the power of individual and collective financial actions in *driving* systemic change.

Money talks. Personal and corporate donations act as seeds, funding research, action campaigns, and pro-environment initiatives that collectively contribute to the reversal of global warming.

Similarly, investments, particularly in green industries, stimulate technological advancements and the implementation of cleaner, climate-friendly solutions.

Government spending and public funding can speed up a shift towards a sustainable future on a large scale.

Overall, each financial contribution, regardless of size, has the potential to shape our planet's future, underscoring the critical role of financial support in our battle against climate change.

You might think that you, as just one person, can't have much of a financial impact, right? But let's view it another way. Think of your contributions as a single ripple created by a pebble tossed into a pond. Although that ripple starts small, it gradually extends ever outward, getting bigger and bigger. That's exactly what your individual contributions can do. A donation made to a grassroots group working hard to protect local forests, a change in your own lifestyle to reduce your carbon footprint, or even having a conversation with someone about climate change — all these actions are like those small pebbles.

That's why even if you can only spare $5 per month, you should. Your monthly donation, combined with monthly donations from thousands of others, adds up to real, *stable* monthly income for the organization. It allows the organization to budget, to plan, and to act. And yes, you should sign up for an automatic monthly plan. Life gets crazy sometimes, and even with the best of intentions, we can forget to send the money in. It's also just less hassle if the payment is automatic.

But Which One?

Finding the right environmental organization to donate to can seem daunting. There are so many! But there are ways to narrow down the choices.

First up, determine your passion. What aspect of the environment resonates with you the most? Is it conservation of forests, protection of endangered species, clean energy, or perhaps soil health? Given that we need to throw everything we've got at our many problems, your donation will be welcome just about anywhere. Once you've got this sorted, start your search for organizations working in that arena.

Now, onto the important stuff. Take a good look at their mission statement and their approach. Are they practical and science-based? Have they achieved measurable results? Check the financials of your shortlisted organizations. Trustworthy organizations have transparent financial information available. They should spend a significant percentage of their funding on their programs instead of administrative costs.

Read reviews, both good and bad, and check their ratings on charity assessment platforms like Charity Watch[1] or Charity Intelligence[2]. Above all, make sure they align with your values and that you feel good about supporting their hard work. In the end, donating to an environmental cause should be as fulfilling for you as it is beneficial for them.

Investing

I touched on adjusting any retirement investments to be greener earlier, but let's look at it a bit more in depth here.

By consciously putting our hard-earned money into things like renewable energy, pollution remediation, or regenerative agriculture, we're not just embracing an economically attractive opportunity, but we're also supporting innovative solutions for a healthier planet. We are encouraging the growth of these industries, helping them to become more affordable and widespread.

Plus, by investing in green energy, you become a part of our renewable future, supporting the transition to cleaner modes of power production, reducing air pollution, and limiting our greenhouse gas emissions. Think of it as double insurance: trying to earn money for your future as well as ensuring we actually have a future.

But Again, Which One?

Keeping in mind that I'm not an investment advisor, nor do I play one on TV... I personally would choose some sort of fund or a set of bonds, rather than placing individual bets on specific projects, companies, or technologies.

Investing in newer projects is inherently more risky than blue chip companies, and funds and bonds will allow you to spread that risk out. Funds are also managed by people who spend their days analyzing the financials of the sectors they invest in, so they're likely to be much more knowledgeable than you or I will ever be.

As for bonds, governments or large corporations usually issue these. At one time, most bonds could be considered sturdy, reliable investments, but given the chaotic era we're in, I regard them as just as risky as funds or stocks.

In any investment situation, be aware that you could lose your investment. Never, as they say, bet the rent money.

If you see an individual project or company doing work that really, *really* interests you, and there is an investment opportunity, take your time and do a lot of research on it. Remember, in any crisis, there will be a lot of scammers and con artists looking to make a short-term buck, and they will be happy to take advantage of your green intentions. Keep your bet small, no matter how promising it seems. Speaking of which...

Beware Greenwashing

When a company spends more time and money claiming to be 'green' through advertising and marketing than implementing sustainable practices, that's a form of greenwashing.

For example, we hear a lot about things like "clean coal" and "renewable natural gas;" these are definitely greenwashing. Clean coal is marginally better than regular coal, but it still produces tons of emissions. The same with natural gas. Indeed, natural gas is arguably worse, because as a greenhouse gas, methane is a much more potent greenhouse gas than carbon dioxide.

These companies can be a trap for investors interested in backing environmentally friendly organizations because the dangers of investing in greenwashing companies are multiple. Not only is the investor misled into believing they're supporting positive environmental actions, but their money also ends up backing harmful practices in disguise.

Financially, these companies can also be a risky investment. Regulatory bodies around the globe are cracking down on greenwashing. As this scrutiny increases, those companies caught exaggerating or falsifying their green credentials can face severe penalties or lawsuits. This regulatory action can result in not only a damaged reputation but also a potential decrease in the value of your investment.

Each dollar directed towards greenwashing companies is a dollar less for sustainable organizations that are actively fighting climate change. So, be vigilant and consumer savvy.

Listening to Indigenous Voices

> *We in the West, with our way of thinking of the natural world, we are not the norm — we're the anomaly.*
>
> — *Wade Davis*

The term 'Indigenous voices' refers to the perspectives, experiences, stories, knowledge, and wisdom of Indigenous peoples. They are distinctive cultural groups associated with particular geographic regions or territories, and they have traditions and historical ties to those places that existed prior to colonization by other societies. The voices of these people carry the richness of their cultural heritage, oral histories, spiritual practices, systems of governance, and especially, deeply ingrained knowledge of the land and environment.

Despite the profound wisdom they possess, these voices have often been marginalized, silenced, or overlooked in mainstream dialogues, particularly regarding matters of environmental and social sustainability. Recognizing and amplifying Indigenous voices is a crucial aspect of reconciliation, and it fosters a more inclusive and holistic understanding of our shared world.

Indigenous knowledge *could* play an important role when it comes to environmental sustainability. Why? Because Indigenous communities have sustained their environment for centuries. Traditional ecological knowledge (TEK) refers to the evolved understanding and skills gained by Indigenous cultures through centuries of direct interaction with the environment. It includes knowledge about the environment, medicinal plants, agriculture, wildlife behaviour, migratory patterns and much more. TEK integrates cultural traditions, holistic views, spirituality, and community wisdom to manage and sustain ecosystems.

Indigenous knowledge also reinforces a worldview that sees humans and the environment as an interconnected web, highlighting the need for harmony, balance, and regeneration, rather than pure resource exploitation. This approach can impact our attitudes and decisions towards the environment and help to drive sustainable change.

You'll notice I titled this chapter with the word *listening*. We have already done terrible harm to Indigenous communities. Colonization and resource exploitation have had devastating impacts on Indigenous populations around the globe. The colonizers, seeking wealth or territorial expansion, often disregarded indigenous rights, leading to forced displacement, land dispossession, and brutal suppression of cultures and languages. This not only disrupted traditional ways of life, but also led to profound loss of local ecological knowledge.

Unchecked exploitation of resources, like logging, mining, and industrial agriculture, has caused severe environmental degradation, affecting the intimate connection Indigenous communities have with their land. These activities have often polluted water sources, depleted soils, and damaged wildlife habitats, affecting the well-being and livelihood of these communities. Imposing western property rights and systems of governance has upset traditional communal management practices, prompting conflict, further impacting the sustainable stewardship of land and resources.

Colonization and resource exploitation have amounted to a double blow: harming both the cultural fabric of these communities and the ecological balance they had nourished for generations.

Indigenous knowledge should *not* be viewed as a resource to exploit to fix the mess that we've made of the planet. Nor should we, in our rush to find the resources we need to build out renewable energy infrastructure, ride roughshod over Indigenous land rights yet again.

It's also important to realize that Indigenous groups are not a monolith. There are as many experiences, opinions, motivations, and agendas as there are individuals. Moreover, Indigenous groups live in a colonized context and are trying to recover from centuries of abuse and oppression. With these two things in mind, it shouldn't surprise you to find out that, for example, while some Indigenous peoples have opposed oil and gas developments (as at Standing Rock), other Indigenous peoples earn a living from the oil and gas industry. Or forestry. Or fishing, and so on.

What does all this mean for you, personally? Begin by educating yourself. Delve into books, documentaries, podcasts, and websites that document Indigenous histories, traditions, and knowledge. Follow and interact with Indigenous activists, scholars, and artists on social media. Check out platforms highlighting Indigenous voices like Indigenous Peoples' Major Group[1] or The Indigenous Environmental Network[2].

Engage with local communities if you have the opportunity. Attend public events, lectures, art exhibitions, or cultural festivals organized by these communities. When you do, step back and *listen*. Learn the nuances. Respect their space and do not impose your views. Ask questions with humility and openness.

Again, don't treat Indigenous wisdom as a resource to extract, but as a perspective to understand and value. Use this understanding to inform your actions, from everyday living to taking part in wider climate and social justice movements. Remember, listening isn't a one-off act, but an ongoing process of growth, respect, and allyship.

Right to Repair

> *Repair is key to ending the model of 'take, make, break, and throw away' that is so harmful to our planet, our health and our economy.*
>
> — *Frans Timmermans*

At its core, the right to repair is a call for regulations and policies that allow consumers and independent repair shops to fix the things they've purchased.

You might be surprised to find out you didn't have that right. But you haven't had it in a while.

Historically, repairing was a common practice! Items were built to last and fixing them was part of their lifecycle. This paradigm shifted in the last maybe fifty years or so, possibly with the advent of modern electronics, where products became more complex and required specialized knowledge. Some people didn't want to take the time to learn what was necessary to make a repair. But at the same time, manufacturers started restricting access to repair manuals and parts, and even started making proprietary specialized fasteners on their products that required special tools to open them.

That, in my view, was a money grab. Restrict the means to repair, make a captive audience for your repair services! Probably doesn't hurt profits if you make sure your product contains parts that wear out fast.

In an age where smartphones, laptops, and other gadgets are ubiquitous, this right influences how we interact with the technology we own. It's not just those devices though! It's everything from tractors to cars to rice cookers.

For consumers, the right to repair is a battle for control over the products they own. It's about having the choice to fix a device without facing exorbitant costs or being forced to purchase a new one. And this freedom is not just about saving money; it's about empowerment and the satisfaction of solving problems independently. It's also about choice.

And it's also about the little shops versus the big corporations. Local repair shops stand to benefit from the right to repair. By having access to necessary tools and information, these small businesses can thrive, contributing to the local economy and creating jobs. This supports community development and provides consumers with more options for repair services. Without the right to repair, money flows to big corporations, typically headquartered somewhere other than your hometown.

Obviously, the environmental argument for the right to repair is also compelling. Electronic waste is one of the fastest-growing waste streams globally, fuelled by the short lifespan of our things and the difficulty of repairing them. By extending the life of our tools and appliances, we can reduce e-waste and other kinds of refuse and minimize our ecological footprint.

As I note elsewhere in this book, I have had to look after three estates, and the sheer number of half-dead kitchen appliances, computer peripherals, and machines that I had to contend with was frustrating. So much "stuff" just taking up space, resources that could be better deployed elsewhere.

There's a lesser-discussed advantage of the right to repair: it *could* drive manufacturers back to designing better, more durable products.

It used to be the case that a washer and dryer would last for 20 years, perhaps more. Fridges, stoves, and dishwashers too. We're lucky now to get three years of use out of them before something goes badly wrong. Finding someone to service your appliance is a huge hassle (because once they have the monopoly on getting your vital device fixed, they don't give a crap about customer service!) and inevitably they don't have the part in stock, and meanwhile your laundry / dishes etc., pile up as you wait. As I've gotten older, I've come to perceive this kind of runaround as time theft.

Of course, manufacturers often argue against the right to repair, citing intellectual property concerns, safety issues, and the potential impact on product quality. While these concerns miiiiight be valid (or might not, as we had the right to repair before and corporations did just fine), they must be balanced against the broader benefits of repairability.

Policy and Progress

The right to repair movement has seen varying levels of support around the world. Some countries and states have introduced legislation to protect this right, because of course this is something that must be legislated into being (rather than corporations doing this because it's the right thing to do). If you want to learn about how it's going in your country, check out https://www.repair.org/[1] or search the term Right to Repair [your country].

Public Transit

> *The failure to invest in our public transportation and public life, I think, is a scandal and a shame, and it should be a national embarrassment.*
>
> — *Mark Shields*

In our collective journey towards environmental sustainability, the role of public transportation *cannot* be overstated.

Public transportation, in its many forms — be it buses, trains, subways, or trams — offers an efficient way to travel. By moving large numbers of people simultaneously, it lowers the *per person* emission rate compared to individual car travel, even when it uses fossil fuel. And of course, advancements in technology are paving the way for even greener options, such as electric buses and electric trains. Or even hydrogen-powered trains, like the one already running in Quebec.

Public transit also reduces the space taken up by transportation. Consider a bus that seats fifty versus fifty separate cars on the highway. Or consider how much parking you need for one bus versus fifty cars.

North America's Ambivalence

Despite these benefits, North America's embrace of public transportation has been lukewarm at best. This attitude stems from a complex tapestry of historical, cultural, and infrastructural factors.

The development of North American cities has, for too long, favoured the automobile. Sprawling urban landscapes with residential areas distant from commercial and industrial hubs make car ownership seem not just a convenience, but a necessity.

There's also a cultural attachment to the personal automobile, as it is often perceived as a symbol of freedom and success. Decades of automotive advertising and a romanticization of the open road, as depicted in literature and cinema, bolstered this perception. An astonishing amount of car ownership is wrapped up in personal identity, status, and 'tribal' affiliations, and yes, even gender roles. Seriously, you could write a whole dissertation on this. I'm sure someone has already.

It doesn't help that chronic underinvestment in public transit infrastructure has led to services that are often perceived as (and sometimes definitely are) unreliable, inefficient, or inadequate. This perception feeds a vicious cycle, where the lack of usage leads to reduced funding and a further decline in service quality.

Charting a New Course

However, the tide is turning. With the increasing urgency of climate change and the obvious benefits of public transportation in reducing our carbon footprint, there is a growing realization of the need for change. Want some talking points? Here are some things we need to do to enhance the appeal and efficiency of public transportation:

- Governments must prioritize funding for public transit to improve service quality, frequency, and reach. This includes investing in newer, cleaner technologies to make rides more environmentally friendly and comfortable.
- Connectivity is a big deal! Transit must be intermodal, easy to figure out, and easy to pay for, to help people get as far as possible in a public system. It should be convenient, not a hassle. Honestly, where public transit works well, it saves you the bother of finding parking, stiff fees for the same, the stress of driving (in bad traffic and in all winds and weather), and the time taken up by driving. Give me a train seat and a good book any day.
- Cities should be designed with public transit in mind. This involves creating transit-oriented developments where residential, commercial, and

recreational spaces are built around transit hubs, making it easier and more efficient to use public transport.

- We need a cultural shift to alter perceptions around public transit. This involves public awareness campaigns, emphasizing the environmental, economic, and social benefits of using public transportation.
- Safety is a big concern about public transit use in North America, because drug addiction, poverty, and untreated mental health conditions are also big concerns in North America. (They're issues everywhere, of course, but compared to Europe, North America does a much worse job on these fronts). Addressing those issues will improve all aspects of life, as well as public transit. Some investment in security in the meantime will also help.
- Encouraging the use of public transit through incentives like reduced fares, tax benefits, and priority access can be effective. Implementing policies that discourage excessive car use, such as congestion charges in city centres, can also be beneficial.
- Engaging communities in the planning and promotion of public transit can ensure that services meet the actual needs of the people. This approach can also foster a sense of ownership and responsibility for public transit systems.

There are many advocacy groups working on improving public transit throughout the world. Here are some notable groups from different regions:

Transportation Alternatives[1] (United States): Based in New York City, this organization advocates for better bicycling, walking, and public transit for a more livable city.

Transport & Environment[2] (Europe): A European-based organization that campaigns for cleaner transport. They focus on reducing the environmental impact of transportation and promoting sustainable mobility solutions.

Public Transport Users Association[3] (Australia): An Australian organization advocating for improvements in public transit in Melbourne and Victoria. They aim to represent the interests of passengers and push for better service quality.

TransitCenter [4](United States): A national foundation based in the U.S. that works to improve public transit and urban mobility. They fund research, advocacy, and innovative policy solutions.

Transport Action Canada[5] (Canada): A national non-governmental organization promoting sustainable transport and smart urban transit policies across Canada.

The Institute for Transportation and Development Policy [6](International): An international organization with a focus on developing bus rapid transit systems, cycling, walking, and non-motorized transport in cities worldwide.

Sustrans[7] (United Kingdom): A UK-based charity enabling people to travel by foot, bike, or public transport for more of the journeys they make every day.

The Eno Center for Transportation[8] (United States): An independent, non-partisan think-tank that promotes policy innovation and provides professional development opportunities across the transportation industry.

Even if you don't have, and are unlikely to get, access to public transit where you live any time soon, advocate for it. Not only will you benefit from reduced emissions and overall pollution, but you'll reduce congestion on the roads.

Labour Action

The essence of trade unionism is social uplift. The labor movement has been the haven for the dispossessed, the despised, the neglected, the downtrodden, the poor.

— A. Philip Randolph

You might not think of unions as being particularly "green," given their long association with industrial work. However, environmental activism and labor exploitation share common roots in fighting against unsustainable, unhealthy, and dangerous practices in the workplace and unchecked corporate power.

Forward-thinking union leaders[1] have recognized that the climate crisis and issues of social justice are as important to their members' well-being as the jobs they seek to protect. Increasingly, they're also recognizing that the economy is transitioning away from fossil fuels and a carbon-based economy, and they're seeking proactive concessions for their members, like skills training to help ease the shift. Because of their power, unions can be an important force in environmentalism. If you're part of a union, you can help shape your local's policies on these matters. Here's how:

Advocate for green jobs: Rather than stubbornly hold the line on old-economy jobs, your collective bargaining unit can advocate for jobs in a greener economy. These include sectors like renewable energy, sustainable agriculture, and eco-friendly manufacturing. They not only contribute to environmental conservation but can also offer decent livelihoods.

Promoting Workplace Sustainability: You can encourage businesses to adopt sustainable practices. This includes waste reduction, energy efficiency, and minimizing carbon footprints, while ensuring that these initiatives are worker-friendly and job-creating.

It should be the case that changes to 'greener' practices are also better for worker health and well-being. For example, the mining industry is transitioning to electrified heavy machinery. Diesel fuel particulate is a huge concern for worker health, and even with (massive and expensive) air filtration systems, miners still breathe a lot in. Electrified equipment is much quieter, which reduces stress and reduces hearing damage. Vibration is also lower, a boon because constant vibration can cause muscle and joint injuries, as well as nerve damage.

Greening Union Facilities: Unions can lead by example by implementing sustainable practices in their own buildings and facilities, such as installing solar panels, implementing recycling programs, and using energy-efficient appliances.

Training and Education: Unions can offer training programs to their members about sustainable practices and the importance of environmental conservation. This education can extend to understanding climate change, energy efficiency, and waste reduction techniques.

Influencing Policy and Legislation: With their political clout, unions can influence environmental policy and legislation. They can lobby for laws that protect the environment, such as stricter emissions standards, renewable energy incentives, and conservation initiatives. These benefit everyone, but especially the workers on the front lines.

Supporting Community Environmental Efforts: Unions often work with charitable groups as part of their community service mandates. Your local can collaborate with environmental groups and initiatives, providing support to help hands, resources, and advocacy.

Environmental Stewardship Programs: Unions can develop and participate in environmental stewardship programs, such as tree planting, clean-up drives, and conservation projects, encouraging member participation.

Encouraging Sustainable Transportation: They can advocate for and support the use of sustainable transportation options for their members, like carpooling, public transport, electric vehicles, and biking incentives. Chargers in the factory parking lot? Why not?

Green Collective Bargaining Agreements: Unions can negotiate collective bargaining agreements that include environmental clauses, such as commitments to reduce greenhouse gas emissions, to use environmentally friendly products, and to incorporate sustainability criteria in decision-making processes.

Population Growth

> *If society will not admit of woman's free development, then society must be remodeled."*
>
> — *Elizabeth Blackwell*

I'd be remiss if I didn't address concerns about population growth. As of this writing, the world's population is at about 8.1 billion people. That's a *very* big number, and certainly the way we live right now, that's not sustainable.

Unfortunately, a lot of the discussions you see online about population growth make some incorrect and frankly racist assumptions. The thinking goes like this: 'those people over there [vague gesturing at a nonwhite developing country] have too many babies and they're a burden on the planet.' This is usually followed up with 'something should be done.'

The name for this kind of thing is ecofascism, which is just plain old fascism wrapped up in paper-thin environmental concerns as a kind of green camouflage. Don't fall for it. Not only is fascism objectively horrendous, the rationale here is just wrong.

On a per person basis, the emissions of folks in North America are [1]*significantly* higher than they are in most developing countries. In 2023, the US put out roughly 14.9 tons per person, India 2.0 tons. Canada put out 14.2 tons per person, Kenya less than 0.5 tons.

On an absolute numbers basis, the highest current emitter in the world is China. But even that's a nuanced number, so don't get too smug if you don't live there. Look at the "Made in..." labels on the stuff in your house, and you'll find that a good chunk of it was manufactured in China, or some other country outside of North America. We essentially outsourced our pollution, as well as the manufacturing jobs, to countries with histories of being desperate for work and a willingness to be lax on environmental and worker protections.

We should also keep cumulative carbon emissions in mind when considering responsibility. A bar 'race chart' by Carbon Brief, shown in a YouTube video[2], suggests that the US is still the all-time leader for total emissions since 1750. (It's embarrassing that my home country of Canada, with its relatively tiny population, is in the top ten on this chart.)

Even worse, these countries with much lower per capita rates are going to bear the brunt of climate change.

Drought, famine, increasingly powerful natural disasters, heat waves... it's going to be (and in many places already is) brutal. All the pain, none of the gain, and most of the blame. It's a terrible trifecta, and part of the reason activists say that climate justice is also social justice.

But let's get back to the issue of population. How *do* we address this without getting into horrific authoritarian policies, fraught with reprehensible moral and social consequences?

There is a significant movement in North America and Europe where people are addressing this by choosing not to have children. They have done so for a variety of very personal reasons, and we should respect their decisions on this. Indeed, it would be so much better if we didn't have so much societal pressure to have kids. Not everyone wants to be a parent, certainly not everyone is cut out to be a parent, and the financial and time resources to be a good parent are tremendous.

We can't count on the voluntarily child-free to save us, however. Fortunately, there is another answer, and one that has several associated benefits: making sure everyone, but especially women and girls, has access to education and reproductive rights.

When empowered, women choose to have fewer children, and have them later in life when they and their partner are more mentally, emotionally, and financially capable of being good parents. The standard of living rises. The incidence of teenaged pregnancies, child poverty, and sexually transmitted diseases falls. It's better for everyone[3] when women have reproductive choices, to say nothing of how it's much better for the individual women.

So, if population growth and poverty and social justice are your issues of concern, you can tackle them all by fighting for women's education and access to reproductive health care. Here again, if you live in North America, don't get smug or complacent about the current level of access. In the US, there are now open, active campaigns against birth control.

Developing Countries

> Beyond the borders of wealthy countries like the United States, in developing countries where most people in the world live, the impacts of climate change are much more deadly, from the growing desertification of Africa to the threats of rising sea levels and the submersion of small island nations.
>
> — Amy Goodman

We touched on how developing countries have not had the chance to enjoy the higher standards of living brought about by industrialization but *are* suffering from the resulting climate change and its worst effects. Coupled with our long history of resource exploitation and colonization of many of these same countries, and, well, I think we can say we owe the citizens of these nations some help.

Climate reparations refers to a call for the so-called "Global North" to help the "Global South" navigate climate change. At the political level, this means money and support. But you don't have to wait for the politicians to get on with it. What could it mean at the personal level?

If you already do so, you can continue donating to organizations that address education, health care, economic development, water and sanitation, and technology access. Those are all very important things.

You can also work to ensure that in our quest to roll out renewable energy that we don't exploit the south all over again. Lithium extraction policies in Chile are already a concern because of the water use involved; labour exploitation is a huge issue in Congo where we source a lot of cobalt. I note elsewhere that these are not valid arguments *against* electrification, renewables, or decarbonizing: these are arguments *for* better, cleaner, and equitable extraction policies for *all our resources*, no matter what they're used for, or who is involved.

Beyond that, however, what can you do?

You can start by sourcing your favourite vices, like coffee or chocolate, from fair trade suppliers. These will be more expensive than other things on the market, but that's the point: the other stuff is cheaper because someone (usually the producer) isn't getting adequately compensated.

Where there isn't a fair-trade option, investigate who you buy other foodstuffs and clothing from. Are they using a sustainable supply chain? Are they a Certified B Corp[1]? Do they track farm to table? What are their supplier labour practices like? And while you're looking at labour practices overseas, don't forget to look at the farmers in your own country. Many use itinerant foreign workers, and the work is backbreaking, done in the blistering sun, and the living conditions for these workers are often terrible. They also typically pay by the piecework system, which means workers have no choice but to skip safety practices to make a liveable wage.

You can also directly fund climate-specific projects. An example would be the Rotary COMMIT [2]project, which aims to provide clean cookstoves. Some three billion people currently cook on open fire stoves, and the smoke contributes to approximately 4 *million* premature deaths every year. The wood for these stoves results in massive deforestation and carbon dioxide emissions, a double whammy. Better stoves use less fuel, burn cleaner, and take smoke away from the family home, improving indoor air quality. To someone living in North America, they're relatively cheap to purchase and donate too.

One of the biggest things, though, is funding and otherwise supporting agriculture developments that focus on empowerment. Many people in developing countries are subsistence farmers. Projects that help them improve their crop yields and reduce their inputs are critical to reducing agriculture's impact on climate change, while also directly improving their lives.

Social Media

> Distracted from distraction by distraction.
>
> — T.S. Eliot

Social media can be its own kind of hell. It consumes hours of our day (if we let it) amplifies the worst and most stupid among us (if we let it) and can be super depressing. There's a reason why we call browsing social media "doomscrolling." If you're not already on social media, it's probably best if you keep it that way.

But if, like a lot of the rest of us, you have been sucked into it for one reason or another, then you can do something to make it better. Rather than being a troll, you can be like the Lithuanian elves[1] who fight disinformation and doom online.

You can start by sharing articles that discuss innovative *solutions* and technologies that address climate change, from reputable sources. There's plenty of bad news out there; the people that are already concerned know this already, and those that cling to their belief in 'climate hoaxes' won't be convinced otherwise. So spread the good news instead. Amplify those people that are taking action. Follow them on whatever platform you favour. Highlight victories and when we make progress. Share, like, engage, and so on.

Share petitions, webinars, and fundraising links to encourage other people to act. Don't be strident, and don't harangue people about it; indeed, most of these things don't even require comment. Ignore anyone who suggests you're "virtue signalling." Once, not so long ago that meant someone who paid lip service to an ideal just to look good; now it's applied to anyone who does anything perceived as "progressive."

If you want to encourage people to eat less meat or no meat, then simply share some of your favourite low and no meat recipes from time to time. Again, no commentary required apart from maybe how tasty it was.

Finally, you can also use social media platforms to organize. I run a Facebook group dedicated to planting pollinator gardens and native plants, you could do something similar wherever you live, on a topic that's near and dear to your heart. If you do start one, keep an eye on it, because a hazard of life online is junk posts and scammers. You'll need to block people from time to time for this reason, and don't hesitate to block or boot anyone who comes into your group just to troll either.

Part 2
At The Individual Level

What you do makes a difference, and you have to decide what kind of difference you want to make.

— Jane Goodall

In an earlier section, we talked about how about 100 companies are responsible for most of the emissions being released right now. We also noted that we as consumers don't really buy from them directly. So how do we fight back at the individual level? How can we curb emissions and cut other kinds of pollution?

By drying up the market for the products that these companies backstop. Voting with your wallet.

I'm not talking about temporary boycotts. Nor am I talking about standing around in the grocery store debating the morality of choosing Brand A over Brand B while the kids are bugging you for chocolate bars or plushies.

What I'm suggesting is a longer-term solution and a systematic approach, whereby you look at one purchasing habit at a time, fix it, and then move on to the next habit.

Let's use an example. Consider this:

Every toothbrush you've ever used in your life is almost certainly *still around.*

Numbers vary, depending on the conditions and type of plastic, but most figures suggest that it takes hundreds of years for plastic to decompose. (And mostly it seems to go into smaller and smaller bits of plastic, which we're now finding in human placenta[1].)

Kind of a scary thought, yes? It gets worse if you do some math.

Let's say the average person changes out their toothbrush once a year.

Now consider a small city with 1,000,000 people in it. That means one city generates a million used toothbrushes every... single... year.

Thus, we're creating *mountains* of used toothbrushes that will still be around for centuries.

And of course, since you're reading this, I probably don't need to tell you that most plastics come from fossil fuels, so this is another market for a product we really shouldn't be using.

So, is there a perfect, zero-impact replacement for plastic toothbrushes?

No. Not yet anyway. But there are products that are arguably *better*.

And indeed, that's the case with nearly everything we'll look at in the coming sections.

Nothing proposed will be perfect, and there's no such thing as zero-impact. Given the size of the human population, the law of big numbers will mean that everything we do will have outsized effects on our planet.

But we no longer have the luxury of waiting for perfect, heaven-sent solutions.

This is so important to note that I'll say it again: we no longer have the luxury of waiting for perfect.

We all must start moving *everything* we do to at least *less bad right now* while en route to those solutions that are sustainable or even regenerative. Too often environmentalists in particular and progressives in general have gotten caught up in arguments and in-fighting over policies and solutions with the result being that nothing gets done.

For example, I would love it if my municipality would increase public transit options where I live! My *nearest* bus stop is a half-hour's walk away up a busy highway with no sidewalk, and no bike lane, so not only not convenient, but dangerous. And the bus has a very limited range and schedule. So, while I campaign for better public transit, I switched to driving electric. I did what I could in the situation I found myself in, rather than waiting for perfect or snarking at people who don't bike or walk everywhere.

You can too.

We'll start with some of the biggest impact items in your household. One will be obvious, and you've probably been thinking about it already. The others might surprise you.

A Note About 'Making Your Own X.'

Before we get into the weeds, so to speak, I want to acknowledge something I see in a lot of online groups related to environmental topics: Yes, it is possible to make your own toothpaste, for example, or mayonnaise, or sew your own clothes, etc., etc.

If that's something that appeals to you, go for it!

But...

Please be aware that doing-it-yourself requires time, talent, good health, money, and in the case of growing food, land. Those aren't available to everyone. Someone living in an apartment is only going to be able to grow a token number of veggies on their balcony. A person living with chronic pain might struggle just to boil pasta for dinner some nights, much less have the energy or focus to make a blouse.

Pffft, I hear some of you say (because I've seen people say this online). Excuses! Why, I held down two part time jobs as a single parent and I still managed to make all my own cleaning solutions, bake all my own bread, walked 12 miles to work every day, canned 456 bushels of tomatoes in a single season...

To which I say, fabulous. Amazing. Seriously, congratulations. It blows my mind what some people are capable of.

But...

Why do we accept that there are only two possible options here? Why does it have to be a choice between planet-destroying industrial production and consumption or going back to subsistence agrarian times where we spent almost all of our existence wearing ourselves to nubs just putting necessities on the table?

Bruce Sterling, on his veridian design website, once called this being "hairshirt-green:"

> Hairshirt-green is the simple-minded inverse of 20th-century consumerism. Like the New Age mystic echo of Judaeo-Christianity, hairshirt-green simply changes the polarity of the dominant culture, without truly challenging it in any effective way. It doesn't do or say anything conceptually novel – nor is it practical, or a working path to a better life.

The dichotomy I describe above strikes me as a failure of imagination. No, we're not going to techno wizard our way out of trouble, but neither do we need to go back to the caves. Surely, if we can manage to put men on the moon, we can figure out a sustainable middle path here.

So, let's stop making the fact that some people can't — or yes, even just don't want to — knit their own mittens into some kind of moral failing, okay? Let's instead normalize demanding better of our *systems*, and using our brains to come up with better ways for *everyone* to live as they choose.

And a Note About Recycling

There are many people who will tell you, with good reason, that recycling is a sham.

For a start, there are a lot of people who don't bother putting recyclable items into a recycle bin. Recyclables need to be sorted, sometimes cleaned, kept in separate bins, and so on. It's more work.

There are also jurisdictions in the world where recycling collection isn't available. In a region where basic water and sewage processing is still hard to come by, you can bet there's not a lot of concern about an old tin can.

Plastic recycling[2] is problematic because the economics simply don't work. They never did. It takes money, time, and energy to recycle plastic. A lot more, in fact, than it takes to churn out new plastic. And plastic isn't infinitely recyclable either, as it degrades. I have seen estimates that suggest that less than 10% of the plastic ever produced has been recycled.

We, the public, recently learned that millions of tonnes of plastic recycling from North America was being shipped overseas to be processed, because the receiver, China, suddenly decided to stop taking it. (Other countries in the region were also taking it in.) Sometimes it would be recycled, sometimes it would be dumped locally, either because some link in the processing chain was corrupt or because there was too much of it. And once again, the rich nations were foisting their problems onto poorer nations.

And so, we're almost literally drowning in the stuff.

That labour and expense associated with recycling, by the way, is downloaded onto you and me, the taxpayers, and the individual, by manufacturers. A soda manufacturer ships bottles of fizzy sugar water out the door, takes the money for the sale, and calls it a day. We get a cheap drink, but then we pay for the costs of dealing with the waste generated. It's the same with every other product.

This reinforces what I've said about the fact that we need major systemic changes. We need container deposit schemes. We need, at the very least, single-use plastic bans and to locate recycling facilities in jurisdictions with decent environmental protection laws. We need to push manufacturers to start closing the loop on the products they make.

In the meantime, we must stop using disposables, cut back on the packaging we routinely receive, and we need to campaign, campaign, campaign for those changes. Keep that in mind as you go through the next part of the book. Wherever possible the first step should be reduce, then reuse, and then, and only then, recycle.

Let's look at how we can do that, room by room.

Outside Your Home

In the Driveway, Garage, or Parking Lot

When we talk about cars — and transportation in general — we have to recognize that the benefits they offer are real, and that the solutions proffered by some environmentalists aren't universally applicable or practical.

For example, the admonition to "just use a bike instead" only works if you:

- Are fit and healthy enough to use a bike
- It's safe and reasonable to use a bike
- You can afford the extra time it takes to get to destinations
- The weather conditions are reasonable

A single mother suffering from asthma, living in a smoggy city with long winters, is not going to be able to cycle to work very often, if at all. (To say nothing of bicycle theft being a very real problem!)

At the other end of the spectrum, driving electric is definitely an option open to more people than ever before. It's still on the expensive side to buy in, though, and again, it's not universally available. And depending on how your local power supply is generated, it might not be the greenest option available... yet.

So, let's go over what you can do to be greener on the transportation front.

Easy and Cheap

The first recourse is simply: do it less! You can use your existing vehicles less often by:

- Grouping errands and trips and not giving into the temptation to just "run out for something."
- Changing up your standard routes. Are there ways to get where you're going that are even slightly shorter? Involve less idling? Remember the law of big numbers. Even if you only cut a couple of minutes off a trip, if it's one you're always making, those few minutes add up to significant savings over time.
- Skipping the drive thru. Unless your vehicle is newer and shuts off to avoid idling, you spew a lot of pollution (and waste a lot of money in gas over time) grabbing that morning coffee. Park and go in, or make something to go at home.
- Take public transit whenever you can — and advocate for more public transit where you live.

- For longer trips, take a train instead of a plane whenever possible.
- And yes, walk or bike if it's feasible.
- Carpool when you can — including with your life partner. Two-vehicle families have become the norm, and you might not need two. Imagine the savings you will enjoy if you can cut back to one vehicle!
- Maintain the car if you can. Simple things like routine oil changes and proper tire pressure can help reduce emissions and give you better mileage.

When it's Time to Replace the Car

- Buy only what you need. Take a hard look at how you use your existing vehicle. How many people do you have in it at any given time? How much cargo do you routinely transport? Where do you drive it? We get sold based on things like off-road capability and towing capacity, but what do we *really* do with our cars? Probably go back and forth — alone — to work and the grocery store. Worse, we pay through the nose for all that unused capacity: in gas bills, insurance bills, maintenance bills, and yes, the environmental cost.
- Remember: you can always rent a vehicle for special needs or trips! I have rented a pickup truck and cargo van for the handful of times in my

life that I've needed those capacities.

- Consider conversion. There's a growing market around converting gas car bodies into electric vehicles.
- Buy used. With stricter quality control processes and inspections for emissions standards, there are fewer "beaters" or "lemons" on the road these days, and you should be able to find a reliable used vehicle. There's no need to continue to fuel the demand for "new, new, new!" while perfectly viable cars sit around in lots. You'll save on the overall purchase price, on interest costs if you're financing, and insurance as well. You won't have to eat the depreciation on a new vehicle either; car resale values plummet by about a third the minute you drive it off the lot.
- Buy based on best mileage for the class.
- Buy hybrid. If electric is not yet an option for you, then do go hybrid, as there's no reason not to these days. Yes, they'll feel a bit different to drive at first, but you'll get used to it soon enough — especially when you can make a tank of gas last for much, much longer. Incidentally, you'll also be able to get better prices for gas because your window to refill is so much longer. You can 'shop around' more easily and not get stung by those price hikes that somehow magically occur right before a long weekend, etc.
- If your old car is on its last legs, consider donating it to an organization that will see it properly recycled in exchange for a tax receipt,

rather than making it someone else's source of pollution.

Please Do Take a Look at Electric

First, in the interest of full disclosure, I personally drive electric, and have done so since 2017. I've racked up 150,000+ km on it so far, and yes that's on the original battery.

If you haven't considered electric before now, you should. And when I say consider, I mean: talk to actual electric vehicle (EV) owners (or lurk in their online forums), read about the vehicle specs, take a few different models for a test drive.

I say this because there is a *lot* of old information and/or deliberate misinformation about EVs online. I get it: new technology can make you feel uncertain, we have a culture that celebrates gas-powered vehicles (vroom!), and oil and gas companies aren't going to go down without a fight. Here's a list of common objections/myths about EVs.

We don't have the charging infrastructure yet

Depending on where you live, you might be surprised. Take a look at Plugshare.com[1] (just close the registration screen that pops up and browse), and look at the Tesla supercharger network[2]. Charging stations are popping up all over the place. Charging stations are appearing at malls, and tourist destinations.

And, in case it's not clear, most EV owners do most of their charging at *home*, overnight. Being able to charge on the road only comes into play if you're taking a road trip or have an extra-long commute. Obviously, this favours homeowners who have a garage or car port or some such, but apartment and condo buildings are increasingly offering charging spots as well.

The range is no good

As of this writing, a Tesla Model 3 has a range of 353 miles or 560 kilometres. The American Driving Survey says that the average person drives about 29 miles a day.

They're no good in the cold

It's true that battery performance is affected by the cold. You can temporarily lose a fair amount of potential range when temperatures plummet. However, given what I noted above about total available range vs. average range driven, there's still a *lot* of margin. If you need to, you can also do things like warm up the car before your trip, recharge at one end, and use the seat heater to stay nice and toasty (rather than more inefficient cabin heat).

But consider this: Norway has the highest market penetration of electric vehicles per capita in the world and has the world's largest plug-in segment market share of new car sales, 74.7% in 2020[3]. Other top markets for EVs include Sweden and Iceland. Not exactly tropical locales.

Oh, and by the way, cold weather also affects gas cars[4]! Your "range" (mileage) is significantly reduced, and in extreme cold you can have issues even starting the car. Water vapour condenses and freezes in the fuel line, oil thickens right up, and the battery that fires the starter can fail. People who live in northern communities have to use a block heater to ensure they have reliable transportation. What about getting fuel? Well, in a recent super cold snap, gas pumps in Calgary froze and no one could use them, so that's a thing too.

All of which is to say that at extreme temperatures affect everyone, we're just more used to how to cope with fossil cars, so it seems okay.

They take too long to charge

For the most part, you will be plugging your vehicle in overnight, or while you're doing something else (like shopping at the mall). On longish road trips, you're going to want to stop for coffee, bathroom breaks, meals and leg stretches anyways. A little bit of planning takes care of both things at once.

The actual charge time will depend on the charger and the car. Older models of both will take longer; my car takes about 30 minutes to get to near total capacity on the supercharger network. Newer generation fast chargers and batteries can get to 80% capacity in about 15 minutes. Cooling cable technology aims to get that down to five minutes.

They're not actually greener

This is a common myth. Although it's true that a new electric vehicle and a new fossil fuel car both (currently) incur carbon footprints to manufacture them, a fossil fuel car will emit pollutants for its whole life. Where the electrical grid is "clean," the electric vehicle doesn't go on to emit *more* pollutants. And recent studies[5] suggest that even cars plugged into 'dirty' grids have better carbon footprints. A battery recycling infrastructure is developing (it's already a multi-billion dollar industry) as more and more EVs are on the road too.

And of course, as the rest of the supply chain electrifies, the carbon footprint of manufacturing EVs will drop as well.

Child labour and mining pollution

In social media comments, you often see claims about child labour with respect to EV battery components like lithium and cobalt. Strangely, these arguments are used to argue against EVs by people who are posting comments with devices that use lithium-ion batteries. Or who are unaware that cobalt has been used in the desulphurization of oil (among many hundreds of other uses) for decades now.

But let's be clear: child labour and pollution are definitely things to be concerned about, and we should of course push international governments to enforce existing laws on such things in *every* industry. Clothing, agriculture, gold mining, even brick production have significant child labour issues that you can campaign against.

EV manufacturers, in the meantime, have made efforts to source their elements from responsible regions, and to eliminate cobalt from their supply chains altogether. Newer battery technology eliminates lithium use too.

They're expensive

The price to performance ratio has been on the high side... until recently. As more and more people have gone EV, the price has come down, while performance has gone up. Prices will continue to drop as the big car manufacturers (finally) get serious about producing EVs.

There may also be government rebates available in your region, so check that out. When calculating cost of ownership, remember that in addition to not needing gas, you also don't need the maintenance that gas cars require — oil changes, muffler fixes, catalytic converter replacement, etc.

The grid won't handle it if we all switch/we need to wait for the grid to be upgraded

It's true that we'll need to upgrade our electrical infrastructure; however, upgrades don't happen without demand pushing them, so 'waiting' isn't really a thing here.

It's also the case that we're going to need to upgrade the grid one way or the other. Climate change is making everything hotter, and we're going to need to keep our indoor spaces cooler. Populations will continue increasing for at least the near future. And climate change fuelled natural disasters are going to disrupt a lot of infrastructure, so building a grid that is far more resilient and has a lot more redundancy is going to be critical over the next couple of decades.

What happens when emergency X happens?

People who like to argue about EVs online have a fondness for invoking extreme situations as the reason why EVs can't possibly work.

One popular scenario is a natural disaster that results in power loss. These are indeed a cause for concern ... for every motorist. Gas pumps run on electricity, so if there's no juice, gas car drivers looking to fill up are out of luck too. And someone who has charging capability at home is far more likely to "have a full tank" in a sudden emergency because topping up is a habit EV drivers get into. Meanwhile, there are connectors you can use that will allow you to use your car as a backup power system for your home in an emergency.

Another popular concern is being stranded in a snowstorm. In a big enough snowstorm, all drivers are equally pooched, and everyone will need digging out and some will need towing. Ironically, because EV drivers don't need to run their engines to run their heaters, they're likely to stay warm for longer, because they're using less energy overall. They also don't have to worry about clearing snow away from the tailpipe to avoid carbon monoxide poisoning.

A more common scenario of concern is being stranded by the side of the road. Again, this is also a concern for fossil car drivers. That's why roadside assistance exists!

Millions of people run out of gas on the road every year and suffer breakdowns if their car is older or not well maintained.

As an EV has fewer points of failure, you're less likely to be stranded due to a breakdown, and you avoid running out of juice by planning ahead, just like you try to avoid the same situation in a fossil car by checking the gas gauge. In the short term, EVs will need towing if they do run out of juice, but in the medium term, we'll build out mobile charging infrastructure in the same way as we have emergency fuel services through towing companies.

The batteries die fast and are super expensive to replace

This is another common claim, and sometimes the contention is that the batteries cost more than the car to replace.

It's simply not true of current EVs. EV batteries are warrantied for 8-10 years, and most owners who were early adopters are finding that even previous generations of batteries are lasting much longer than that.

Once your battery needs replacing (and these head for recycling plants), your cost to replace will be equivalent or less than the cost of a small used car, and you'll effectively get a whole new car out of the purchase. And that's at battery prices at the time of writing. Again, costs will go down as big manufacturers get into the game. (If that still sounds shocking to you, price wise, go price out a replacement gas engine.)

The government wants us all in EVs so they can control where we go!

This is a weird one, but surprisingly common (thanks, Internet). The thinking here seems to be either that you can only go on certain routes in EVs (not true), or that EVs can somehow be taken over by mysterious government forces. Given that fossil cars are also loaded with onboard computers and GPS tracking systems, if the government wanted to stop you from driving via hacking, they could do that now.

A corollary to this one is that an electromagnetic pulse (EMP) attack would prevent you from driving. Again, in this situation, fossil cars would also be disabled by an EMP attack, as would every other computerized system in the vicinity.

In short, there's nothing inherently more "controllable" about an EV.

Fires

Lots of people like to claim that EVs are dangerous because they're a fire hazard. The media doesn't help here, making a big deal out of every EV fire that happens.

But realistically, cars in general are a fire hazard; indeed, gas car fires are so commonplace they're a staple of action movies. We routinely carry large tanks of flammable liquid around in conditions where collisions are high probability and think nothing of it. Gas car fires can also happen spontaneously when parts malfunction.

Both Jeep[6] and Kia[7] announced recalls for fossil cars because of this very issue.

And finally, there is some evidence to suggest (based on an insurance company report) that of the three types of vehicles available to consumers, hybrids are statistically more likely to catch fire, and EVs are least likely with gas cars in the middle.

What *is* true is that fire departments will need different equipment and training to handle EV fires, and that we should be demanding that EV manufacturers work to provide additional safety layers to minimize hazards.

But hydrogen!

Hydrogen powered vehicles will probably be part of our future too. As of right now, it looks like hydrogen will be more practical for bigger vehicles and transportation systems. (I've had the pleasure of riding a hydrogen-powered train, and it was great.) At this moment, there isn't enough infrastructure for the average driver to consider switching. That may change rapidly, or it may take years.

However, given the very, very short runway we have to decarbonize our economy, I would suggest you switch to EV now, and if hydrogen becomes viable in the future and it works for you, switch again then. The goal here is to stop dumping CO2 into the atmosphere as quickly as possible.

You can't build EVs without gas and oil, therefore it's a sham

Oy. This one is just annoying.

In a *transition* period, like the one we're in, you still have to use the old system to produce the new system. Much like the first car factories and rail lines were built with the help of horses. As we electrify more and more of our industrial processes (including the heavy machinery for mining), the carbon footprint of the whole chain will drop. Enough said.

One other thing to consider

As of 2020, it was possible to do a cannonball run[8] across the entire span of Canada (the world's second largest country), in the winter, in a Tesla Model 3. They did it in about 73 hours, too. Total distance was 6131 km (3810 miles).

The first takeaway here is that in 2020 the technology and infrastructure already existed for that to happen.

The second takeaway is that, on the EV front at least, things have only gotten better since then, and will continue to get better in the years to come.

People tend to think that the way things are now are the way they always will be, even when history has shown us, time and time again, that is not true. As I mentioned above, there's already research being done to reduce charge time. There's a lot of money being thrown into research and development to make batteries lighter and more energy dense, and to make use of elements like sodium and sulphur to further mitigate resource extraction issues. This is good news for EVs, but also for energy production, where storage solutions will be needed for renewable electricity systems.

It also pays to look at history. The first gas cars were expensive toys for the very rich. They had terrible range, were prone to breaking down, were difficult to start, and we certainly didn't have gas stations on every corner. Indeed, even now, there are parts of North America where there isn't a gas station for hundreds of kilometres.

People in the horse industry — which had powered our economies for *centuries* — confidently predicted that cars would not last. They happily pointed out all the problems gas cars had.

Yet we pivoted from majority horse use to majority car use for personal transportation in a span of less than 40 years. Change happens slowly and then it happens all at once.

Let's move on to a car *related* environmental concern.

Clearing Snow and Ice

Those of us who live in climates with significant winters know that clearing and de-icing our walking and driving surfaces is a necessary evil.

Fortunately, just like there are now electric lawnmowers, we finally have electric snowblowers and electric snow shovels available. Bonus: push button start! (Ask me how much, as a short person with shorter arms, I like those pull cords for gas engines?!) Read the reviews before you buy, as these are new to the consumer market, and you'll want one that is reliable and powerful. Also try to get the same brand as your electric mower and other yard tools, because the batteries (should be) swappable.

As for de-icing, since the 1930s, we've used rock salt, because it's super effective and cheap. It's also not great for the environment, not in the quantities we use it. It kills plant life near the roads, leaches into the water supply and kills marine life that isn't salt water adapted, and it attracts wildlife to roadsides (they like to lick it up), increasing the probability of road accidents. The stuff is also corrosive and costs us a fortune in road repairs.

So far, there isn't a silver bullet replacement for municipal road crews (alternatives are more expensive, less effective, or come with their own issues when used at scale) but at home, you can switch to an environmentally friendly mix available from your local hardware store. If you want to make your own, alfalfa meal, coffee grounds, sugar beet juice, sand, and fireplace ash are all alternatives. Just be aware of the pets in your neighbourhood and potential consequences if consumed. (As a dog owner, I can assure you that dogs are stupid enough to eat coffee grounds if available, for example).

Lawn

Let's talk about lawns.

The lawn as we know it is a recent invention. You can find its origins in the manicured estates of the European aristocracy. Devoting acres of land to high maintenance, useless grass was a way to flaunt your wealth. It meant you had so much land that you could take some out of crop production and devote it to being decorative. And further, that you had the staff to maintain it to exacting standards.

The standard suburban turf grass lawn — green, neat, tidy — is still a status symbol. It implies that the owner is comfortably middle class or higher. It still requires staff to maintain it too: many a weekend is devoted to trimming, weeding, rolling, spraying, aerating, raking, watering, and fertilizing. If you are especially wealthy, you can afford specialized equipment for the job, or you can have a service do all this work for you.

The lawn as a concept is so ubiquitous that some 128,000 square kilometres are devoted to it in the USA. According to NASA, that's three times more acres of lawns than irrigated corn.

Here's the thing, though: from an ecological standpoint, that green lawn is a desert. It's also a huge contributor to carbon emissions.

Let's address the first point. We clip our lawns very short, so they don't provide a decent habitat for anything but the smallest of ground dwelling insects. We're careful to roll everything flat and fill in holes, so nothing dares burrow. We spray for "weeds" and eliminate any flowers for pollinators. We also don't fancy bugs very much, so we kill grubs and other critters with pesticides. So, from a wildlife point of view, that's tens of thousands of square kilometres that are essentially no longer habitable or food bearing. Combine this with everything we've paved over, and well, there's not much left for anything else, is there?

As for carbon emissions, we can look at direct and indirect production. Most of us still use gas lawnmowers (and gas-powered weed whackers and leaf blowers), which are highly inefficient. The US Environmental Protection agency (EPA) estimates that hour-for-hour, gas-powered lawn mowers produce 11 times as much pollution as a new car. The government of Canada suggests that a single lawn mower produces 48 kilograms of greenhouse gas in a season.

Oh, and they're noisy and smelly! How many Saturday afternoons would be improved by the absence of loud two-stroke engines?

Meanwhile, think of all the emissions generated to bring you: grass seed, weed killer, bug killer, and all the tools and machinery you use. While we're at it, let's think about how wasteful it is for every single household to have a dedicated lawn mowing machine. Several lawnmower maintenance sites suggest that the average lawnmower is used for about 60 hours per year. That means it just sits around in your garage for 99.4% of the year.

Lawns are also water hogs. Landscape irrigation (lawn watering) is estimated to account for nearly 1/3 of residential water use, or 27 billion litres every day (EPA).

Finally, all those lawn chemicals contribute to waterway pollution and fertilizer runoff, which creates algae blooms that kill wildlife.

So, there's very little to love about lawns.

All of that said, it's easy to see why we still have them. They're 'normal' to us because we grew up with them. Kids and dogs like playing on them. And in areas where certain insects — like ticks, for instance — are problematic and bring disease, there are good reasons for keeping nature at arm's length. Plus, we like imposing order on our surroundings.

How to Make our Lawns 'Greener'

- Fortunately, there are a lot of easy fixes, many of which involve *less* work than what you're doing.
- If you're constrained by local bylaws or a homeowner's association and can't (easily) change the composition of your lawn, you can still decarbonize by switching to electric mowers and tools. The latest generation of tools have enough power and torque to handle most lawns and there are even decent electric riding lawnmower options now. You could also shop around for a 'green' lawn service that uses electric machines. If you have a small lawn, you could also just use a reel mower.
- On that note, you can call around to your local lawn service companies and ask if they use electric equipment (even if you already know they don't). If they feel they're losing potential customers by using fossil fuels, they might start decarbonizing their fleets.
- Mow higher. It doesn't need to be shaved down to a nub. It will be more resilient to drought, less likely to get weeds, and will look more lush.
- You can reduce or eliminate your fertilizer needs by adding nitrogen fixing plants to your lawn. Clover, for example, used to be standard in lawns until weed management chemicals killed them off.

- You can also use your compost to fertilize your lawn; once it's thoroughly composted (so you don't attract unwelcome rodents), use it as a top dressing for your lawn.
- Speaking of which, consider accepting imperfection, and give the weed killers a miss. If you must control weeds, switch to corn gluten application in the spring to stop weeds from sprouting in the first place.
- Get your lawn tested and investigate what local mycorrhizal fungi you should add to your lawn. These fungi have a symbiotic relationship with certain plant roots, and help plants absorb phosphorous, potassium, calcium, copper, and iron. They also help with water uptake.
- You can reduce or eliminate your water bill by setting up rain barrels to catch rainwater and save it for dry days. You can also reduce evaporation by watering very early in the morning or late at night when it's cooler. More water gets to the plants this way.
- Other water saving options include setting up 'rain gardens' and 'grey water systems.' A rain garden uses roof run-off and rainwater to nourish water loving plants like lilies and reeds. (Always use what's native to your region.) A grey water system reuses water from things like showers and sinks. Be sure you're using environmentally friendly soaps and not dumping anything bad down the drain.

- You can also just let your grass go brown in peak summer heat. Grasses have a natural dormancy cycle they use to conserve nutrients, and they can stay that way for as much as a month.
- If you have looser local regulations, consider over seeding with a regionally appropriate “low mow” or “no mow” seed mixture. These produce short grasses and flowers that need only a few lawnmower passes per year, or sometimes none at all.
- You can reduce the total area of lawn you need to mow by planting native shrubs and trees.
- You can also convert sections of your lawn to native flower gardens. Or go all out and convert everything! It doesn’t have to be a wild meadow — it’s possible to have an orderly wildflower garden if that suits you more.
- Finally, you might get past local regulations by converting everything to food production. While this isn’t ideal from an ecology point of view, it does save on greenhouse emissions, and you get produce out of it, which will help you save on your grocery bill.
- For the love of all that is holy, do not ‘fix’ the problem of lawns by paving it over or installing fake grass. Concrete production produces CO2, and paving everything contributes to flooding because the water can’t just soak away into the ground. Fake grass is plastic, which is terrible for the environment in all kinds of ways.

Garden

We touched on gardens in the last section, but let's expand on this a bit, because they're hugely important to fixing what ails us.

Drive through any city, and you'll see suburban houses with neat and tidy gardens out front, probably a decent number of tree-lined streets, and maybe some veggie plots. All looks "green" in the environmentalist sense, right?

Possibly not.

There are several problems with modern urban landscaping, and these issues are contributing to — not improving — habitat degradation, and biodiversity loss.

Let's start with habitat.

(And yes, this includes those plants you have on your balcony.)

Garden centres, nurseries, and landscaping contractors sell ornamental plants based on two factors: how easy they are to look after, and how pretty they are.

But unless those plants are 'native' to your region, they're almost useless to your local wildlife.

(As is your veggie garden because you work to keep those "pest" free. Which is reasonable, because you need to eat, so let's keep this about ornamental plants.)

Butterflies, for example, require certain host plants on which to lay their eggs and to eat as caterpillars. As adults, they need other plants to feed on.

Now, you might be thinking that no bugs is not such a bad thing, because who wants a lot of creepy crawlies around? And bug-chewed plants aren't super attractive.

Well, all the other plants that need pollinators (including food crops!) want those bugs around. As do all the birds, frogs, toads, and other insects that eat those bugs. A single nest of songbirds, for example, needs thousands of caterpillars to feed their young. And certainly, all the creatures that eat birds, frogs, toads, and bigger insects need something to dine on too.

To put it another way, imagine if you were surrounded by buffets that either had empty trays, or were only filled with things that you were allergic to (or just can't stand to eat). That's what your garden looks like to the wildlife if doesn't have any native plants in it.

Worse, some of the plants that are sold in stores can be 'invasive.' That is, they escape the garden and get established in the wild. There, they can sometimes flourish, outcompeting native plants and destroying habitat. They do this by seeds (which can be relocated miles way by birds or wind); cuttings or through a weed pull where the debris isn't baked by the sun or composted into mush; through 'runners;' or rhizomes. So, even if that invasive plant looks like it's behaving itself and staying put, it probably isn't.

Which brings us to biodiversity loss.

If we're losing bugs and the things that feed on them, we're reducing the biodiversity of the ecosystem. What that means in plain English is that the entire system becomes a lot more fragile. When you have *lots* of different bugs, and plants, and animals, and birds, one disease, or one food source reduction isn't going to turn the whole area into a desert for you too. The fewer creatures you have in your system, the more vulnerable the whole thing is.

You Can Do Something About This!

That's the bad news. The good news? This is something you can definitely fix.

Let's go back to your garden. Figure out what you have already, either by asking gardener friends (in person or online), snapping pix with the iNaturalist app or using Google Lens, and getting identifications. (You might also have receipts you can look at, or you could ask for a local landscaping company to come and do an audit.)

If you have anything that identified as *invasive* in your area, target that stuff immediately. Pull it up, cut it down, stick everything in black garbage bags, do the bags up, and leave them in the sun to cook for several days. (Be sure to clean up seeds as much as possible). This is called solarizing. You want to make sure there isn't any live material or seeds or cuttings that could spread once you dispose of it. Once it's *thoroughly* cooked and completely dead, you could compost it, or if you must, put it in the trash.

Next, target anything else that isn't native and consider replacing it. Or filling in blank spots with native plants. Or ideally, both! Here's where you can make use of free resources like your local library, your local state or provincial environmental departments, gardening groups, and horticultural societies. Or you can even do a search online for the term "plants (or trees, or shrubs) native to my area." In some districts, enthusiasts have started "plant this, not that" lists, which give you similar looking plants to the ones you want to replace.

You don't have to do everything all at once, of course, as time and money will be considerations. But here again, check out the free resources. Many libraries have started "seed libraries," where you can "borrow" seeds in the spring to start plants and then you harvest seeds from your plant to take back to the library. There are almost certainly plant or seed exchanges where you live too.

A few words of caution. Be careful of plants labelled "pollinator friendly" in nurseries or big box store garden centres, because they might not be what you really need. For example, "butterfly bush" does attract and feed butterflies, but it's from Asia, and has been declared invasive in several regions of North America. Beware too of anything that has a special marketing name in quotes after the flower name. A fictitious example might be Lanceleaf Coreopsis "Summer Blast." This means it's been hybridized and isn't the original native plant. Symbols like ™ and ® anywhere near the name mean the same thing. It might be pretty, but it will be bred for traits appealing to humans, not the critters we need to look after.

I always take a list of what I want to buy with me, using those hard-to-remember Latin names, so I know I'm getting the real deal. There are also now nurseries specializing in native plants, so check their reputation and then you can shop for everything there with confidence.

And, if you have pets, make sure you're not planting anything that would be toxic to them in places they access. Your veterinarian can point you to a list of things that are bad for cats and dogs.

Special Garden Types

Extra dry

If you live in an area that's normally hot and dry, you might need to rethink what 'garden' looks like.

Xeriscaping is the practice of landscaping to reduce or eliminate the need for irrigation. This means using plants categorized as xerophytes; they're adapted to hot and arid climates already. So instead of an English cottage garden full of thirsty hydrangea, you might instead have cactus!

Extra wet

Have you got a spot on your property that's always soggy? Do you perhaps have to deal with a lot of runoff because when it rains, it pours? Live in an area prone to flooding? Consider a rain garden.

A rain garden collects rainwater, holds it for a while, and filters it before slowly releasing the water into the ground. As you can imagine, this helps reduce water flow, prevents erosion, and, if you plant native, provides habitat for all kinds of critters. The filtration aspect also prevents a lot of material getting into our water systems.

There are plenty of resources online about how to build one, including some that can be linked to your gutter downspouts. Bonus: rain gardens might also reduce the chance of water getting into your basement.

Rain barrels

While we're on the topic of controlling water flow and water conservation, consider adding rain barrels to your house. These are big containers that you hook into your gutter and downspout system (or alternatively, place out in the open with collector funnels) to capture and save rainwater. In addition to saving water for a (not) rainy day, which can be used to water gardens, you get some of the same benefits as you get from rain gardens above: water flow mitigation, and erosion control. Note: it's best to use water from roof runoff for flower gardens rather than food gardens.

No till gardening

It's one of those things we do because our grandparents did it: turning over the soil (tilling). It's supposed to break up heavy soils like clay, help you work in soil amendments, reduce weeds, and it will quickly provide a tidy looking field of earth ready to plant.

There's evidence to suggest, though, that the practice might not be the best idea. It increases compaction over time, disturbs the bacteria and fungi systems underneath the surface, brings up dormant weed seeds, rocks, and roots, and reduces the amount of moisture the soil holds. Over time, constant tillage also reduces fertility, which means you must keep adding amendments and fertilizer.

By planting cover crops or plants with deep root systems, you can achieve some of the same benefits of tillage without the drawbacks, just at a slower pace. Bonus: no gas guzzling rototillers needed, and better carbon sequestration in the soil. It's worth checking out how to use no till methods in your garden.

For peat's sake

Another thing we do without thinking: add peat as a soil amendment. It helps soil retain moisture, prevents compaction, it's sterile so doesn't add stuff you don't want, and it's acidic which is great for acid loving plants. Plus, it's biodegradable, so what's not to like?

The problem is that peat moss harvesting is not sustainable. Peat bogs take centuries to form as they grow at a rate of a millimetre per year, according to Monty Don[1], the UK's best known garden writer and broadcaster. Many plants and animals depend on peat bogs, which are unique ecosystems, so essentially, you're destroying very hard to restore habitat to build your own.

Coconut coir is a good alternative. Coir is made from the shell and outer covering of coconuts, and so it makes use of what would otherwise be a waste product from coconut consumption.

Join a Group Doing This Work

You don't have to do this all by yourself either. In Canada, the David Suzuki Foundation has a Butterflyway Project with resource materials and volunteer coordinators set up to encourage more native gardens. The Rotary organization, which is a worldwide service club, also has a pollinator garden initiative. Why not meet new people and make new friends while doing a good deed? See what's available where you live.

Tell Your Neighbours

Talk up native plant gardening to your neighbours. If you've joined a group like I suggested above, there are often signs you can get from them to put on your lawn to explain the initiative and start conversations. They also provide flyers you can surreptitiously (or openly!) put in your neighbour's mailboxes. Spread the word!

What About My City Plantings?

This is something you can influence too. Get together with existing gardening and horticultural groups and start pressuring city hall to adopt more "native flora" planting policies. As always, see if you can find a way to make it about tax savings, property values, and quality of life. You don't have to like the system, but you can learn to work it!

Corporate and Institutional Plantings

Don't forget to target corporate or institutional plantings as well. Think of all those parking lot boulevards, apartment building planter boxes, large acreages around utility company buildings... the list goes on and on. A letter writing campaign can work wonders!

You could also involve your local Scouts and Guides troops and encourage local schools to start pollinator gardens as teaching projects. Consider reaching out to local Indigenous groups while you're at it. Working together to bring back habitat can be a wonderful reconciliation activity, particularly if you spend a lot of time listening and learning.

Don't Forget the Trees

This wouldn't be a proper book on the environment without a section on trees, so let's get to it.

If you're reading this, chances are you already know that trees are super important to carbon sequestration. They take up lots of CO^2 and give us oxygen. (Phytoplankton do this at scale too but let's leave them to the marine biologists.)

What you might not know is that certain trees are keystone plants for pollinators. In North America, for example, native oak trees host hundreds of different types of caterpillars. Trees also provide food and shelter for dozens of other creatures big and small.

You know that trees provide shade, but you might not realize that tree shade reduces what's called the "urban heat island effect," which is a fancy way of saying that our cemented over cities bake our brains in the summertime if we let them. Tree shade reduces the need for air conditioning, so they're a great addition to passive house cooling systems.

Trees help clean the air, by filtering particulates out, and clean the water by removing pollutants and sediments from rainfall and then releasing the water back into the ground or air. Trees are even good for our mental health. Who doesn't enjoy sitting under a tree at lunch? A walk through the woods is even better; the effect is called forest bathing.

Finally, our woody friends control erosion and can reduce flooding, thanks to their massive root systems and thirsty nature.

So yes, please plant more trees. As many as you can fit (and remember, real forests are crowded!). If you aren't blessed with a yard, then please donate to afforestation or rewilding projects. You can also change your default search engine on all of your devices to a tree planting one by using Ecosia[2].

Compost

Tumble and Solar

Another thing we can do to help the environment is compost our food waste.

Every load of garbage we send to a landfill not only doesn't go away (it just sits there for future generations to deal with), but it generates emissions. Garbage trucks haven't yet been electrified to any great degree.

Further, food (and other organic materials) waste produces greenhouse gases. During the decomposition process, food waste that's sealed off in a landfill undergoes *anaerobic* digestion. Basically, when organic material is trapped in a place without oxygen, different types of bacteria break it down. This process releases methane, an even more powerful greenhouse gas than carbon dioxide.

While it's true that we can tap landfills to capture and burn methane, it's not an ideal solution. Landfills leak methane like crazy, and as I've already mentioned, they just punt the problem of waste into the future. The goal should be to stop sending crap to landfills.

You can divert your organic waste right now by composting. Despite what you might have heard, it's quite easy to do.

In our household, we have a simple stainless-steel bucket with a lid on our counter. Vegetable scraps and the odd paper towel that we use go in it, and when it gets full, we take it outside to a tumble composter. Ours has two chambers, so new stuff goes in one side when the other side gets three-quarters full. Every time we add new material, we spin or tumble the composter; that helps aerate the compost, so it breaks down faster and produces less methane. The composter is in a sunny spot, so it gets the full heat of the day to break down the material too. Once or twice a season, we open the 'finished' side of the tumbler, dump out the new soil, and spread it in our flower garden.

With consistent emptying, the kitchen bucket never gets smelly. We scrape it down once or twice a month and hose it out to keep it clean. The compost tumbler outside is a closed system, so we don't get vermin looking for food. You can find a list of what can go in a standard composter here[1].

If you want to expand what you compost, there are a couple of ways to go.

Yard waste can go in a larger compost heap or bin, although ideally, you're not producing much of that anyway. (Autumn leaves help shelter insects and critters over the winter, so we leave ours where they fall until the grass starts growing again in the spring. Then we use a mulching mower to return the material back to the earth directly; same with grass clippings.) Every month, you'll want to take a pitchfork to the pile to allow oxygen in to break things down faster.

You can also use a solar digester. The green cone version we have allows us to handle tougher organics like bones, grease and oils, pet waste, and dairy. It's a bit of work to install — you'll need to dig a small hole in a sunny location — but because it has double wall construction and high heat retention, it breaks things down super fast. And it rarely needs either stirring or emptying; the hole you dug will be for a flow-through basket that allows the nutrients to wash away into the soil.

Bokashi, Vermicompost, and Indoor Composters

There are other options for those of you who don't have a lawn or garden. Bokashi is a Japanese method of fermenting compost. It's good for digesting just about anything, but requires a bit more maintenance, and it's an anaerobic process. The "tea" that is produced is good for watering gardens with.

Vermicompost is where you set up a system to feed worms, usually red wigglers. Basically, the worms eat what you put in and break it down, as well as produce "vermicast" which is worm poop by any other name. It too requires a bit more maintenance than the methods above but has the advantage of being compact and it breaks down the compost fast.

If fermentation and worms aren't your thing, there are an increasing number of "indoor" composting systems available. These use a combination of mechanical tumbling (or grinding or slicing or all of these) plus heat from a warming coil, to break down compost in an appliance. This is a new consumer product, so buyer beware: read the reviews! Some systems advertising "compost" just produce dehydrated chopped up organic matter that might go moldy when spread on your lawn or garden. Others might require you to purchase additives regularly, the delivery of which increases the carbon footprint of the endeavour.

Pets

We love pets. They are a source of joy and companionship to millions of people around the world, our household included.

But we need to start having some serious and honest discussions about their impact so we can work to reduce it.

First, let's talk about the waste that our pets generate. The most obvious source of waste is piddle and poop, and boy howdy is there a lot of it. One online calculator suggests that two smallish (~40 pound) dogs produce something like 500 lbs of poop every year. Now multiply that by the number of dogs there are in your country and, well... holy crap!

Then there's all the waste produced by what we use to handle... all that waste. Kitty litter. Hamster bedding. Plastic scoops. Poop bags.

And what about toys? We routinely hand dogs and cats toys to destroy, which is a huge source of waste. Not to mention all the packaging that the toys come in. Plus, the packaging for animal feed, all the packaging for their medicines, and alllll the resources used to manufacture and distribute the food, the toys, the meds, and things like leashes and collars, cages, and carriers, etc.

Add to this the resources required to handle strays and dumped pets. As well as the resources required to see a beloved animal across the rainbow bridge and deal with their earthly remains.

And finally, the direct environmental damage done by our more mobile pets. Cats, for example, are estimated to kill[1] more than 2.4 *billion* birds a year in the US alone.

It's... a lot.

And it's not sustainable. Not the way we currently do things.

Doing it Better

There are a few ways you can make this situation better, right now.

First, if you don't have a pet already and you're thinking of getting one, please be sure you can commit to it for its whole life. Be sure you're going to be around a lot for your pet as well. We know it's nice to come home to a pet companion, but is it fair to leave a social animal home alone for 80% to 90% of its life? (Alternatively, you could use your pet as an excuse to request remote work; see the section on labour action!)

Please also get a pet from a shelter, which are always full to bursting with all kinds of animals in need of a 'forever home.' This will help reduce the burden that already exists. Make sure you spay or neuter your pet, so we don't increase the population further if your friend escapes for a while.

Source your toys and accessories second hand (there's lots of pet safe stuff in online marketplaces), and let's stop buying new stuffed toys for pooches and cats to shred. If you don't have a second-hand source, look for brands that produce eco-friendly pet accessories (beware of greenwashed claims, as always.) Talk to your vet about safe, sustainable chewing alternatives.

Make sure your kitty litter and small critter bedding is compostable; at our house, a solar digester takes care of pet waste so we're not sending it to the landfill. When we walk the dogs, we use biodegradable bags for pickup, then they go to the digester too. If you don't compost (see the previous section), then at the very least get the compostable bags so the waste will break down faster.

While chonky pet photos might be funny, overfeeding increases the resource burden and is unhealthy for your animals too, which in turn increases their veterinary needs. If you can't resist those cute little eyes, break up a treat into many small pieces to be handed out over time, so their overall calorie intake is reduced.

Speaking of food, buy yours in bulk to reduce packaging, and choose a sustainable brand. Recycle the packaging you do get. If there's a local pet food bakery or manufacturer, source from them.

Keep those cats indoors! If you must take them out, harness train them and keep them on a lead. But don't let them loose; as I mentioned above, they are devastating the bird populations.

Finally, consider avoiding 'exotic' pets. As cool as those reptiles, snakes, spiders, big cats, tropical fish, and colourful birds can be, they're often sourced and traded illegally, which depletes natural populations. Some also need special, energy intensive habitats (like heating lamps or water filters), more intensive food needs (rodents, bugs), and their very real social and exercise needs that aren't being met when they're kept alone in a small pen.

And worse, when these critters escape (and they do), they can be problematic. Koi and goldfish, for example, have become invasive pests worldwide; they are voracious eaters and make waterways turbid and unattractive to other species. Other creatures can spread parasites and diseases that local fauna aren't resistant to.

Power

In recent years, we've reached an important tipping point: the cost of residential renewable energy has plummeted, even while efficiency has increased.

That means a system might be within reach for you, financially. If it is, I encourage you to get one.

Residential solar and wind will be key to reducing the carbon footprint of houses. As this fantastic video shows[1], our use of fossil fuels for both electricity production and driving is not sustainable.

It will also be what gives us *real* energy independence, and a resilient grid.

What do I mean by that?

At the geopolitical level, energy independence means not having to depend on imports for critical energy needs. Even if two nations are friendly, and likely to remain so, all kinds of things can impact the price and reliability of the supply from outside your country.

At the personal level, energy independence means not having to depend on big corporations or utilities for our energy needs. For too long, we've been at the mercy of organizations like OPEC. Oil producers get together to artificially restrict supply, the price at the pumps goes up, we still must go to work, and so we have to pay the going rate. Hurricanes along the US south coast shuts down refineries? We pay. War in Eastern Europe? We pay. You get the idea.

Energy independence also means having some security and stability in your personal power supply. Locally, transformers fail all the time. Accidents can take out the power. Storms can knock out the juice, too. Of course, your own equipment won't be immune to failure or storm damage, but having your own means of producing electricity means you have a backup system that also can be used to supply your daily needs (as opposed to a diesel generator, which is only useful to homeowners in an emergency).

A resilient grid means that we're not dependent on one or two big producers for all our needs. We saw in the pandemic what happened when certain monopolistic industries were hit by illness: shortages, prices skyrocketing, etc.

Energy is mission critical for everyone. We can't mine coal or refine oil in our backyards, but we can produce electricity.

Solar

The first step in getting a solar power system is to evaluate your home's solar viability. Factors such as the roof's orientation, angle, and shading can impact the efficiency of a solar power system. Homes with south-facing roofs without significant shading from trees or buildings are ideal. Obviously, someone in a sunny climate will do best, but, contrary to popular belief, solar panels can still generate power on cloudy days!

Residential solar systems come in two types: grid-tied and off-grid. Grid-tied systems are connected to the local utility, allowing homeowners to feed excess electricity back into the grid (receiving credit through net metering) and draw power when needed. Off-grid systems, on the other hand, operate independently from the utility grid and require battery storage to supply power during nighttime or cloudy days.

The initial cost of installation can be a significant investment, but fortunately, there are now plenty of financial incentives from federal, regional, and municipal governments. Don't forget to price in the long-term savings on electricity bills. The payback period—the time it takes for the savings to equal the initial investment—varies but can be as short as 5 to 10 years, depending on your location and electricity rates.

Before you contact an installer, take some time to do some research on what kind of system you want, what the prices are locally, and what kind of equipment and installation guarantees are available. As with any new thing on the market, there will be lots of legitimate businesses, but also more than a few scammers. Don't be bamboozled. Do your homework first!

Solar power systems are low maintenance, but regular checks can ensure optimal performance. Ask your installer about maintenance services and self-check procedures.

Wind Power

When we think of wind power, you might have those huge offshore turbines in mind. But did you know residential wind turbines are available? Technological and design advances mean that there might be options available for where you live.

As with solar, the first step is to assess your property. To maximize your output, you must have reasonable local wind speeds, and adequate unsheltered space. By that I mean some area where you could mount a turbine (on a pole or on your roof) that can catch a decent breeze. You might be surprised as to how much air movement there is around your house; when I installed a mini weather station last year in an area that was semi-sheltered at a distance by trees, I was shocked to see how much and how often the little anemometer spun.

The next step is to check your local zoning laws and if you're in one of those homeowner associations, their rules too. There might be restrictions on turbine height, noise levels, and location.

HAWT or Not?

Residential wind turbines come in various sizes and designs, with the two main types being horizontal-axis wind turbines (HAWTs) and vertical-axis wind turbines (VAWTs). HAWTs are the traditional windmill-style turbines. They are more efficient than VAWTs and are best suited for areas with higher wind speeds. VAWTs have blades that revolve around a vertical axis. They are more versatile in terms of placement, can operate in lower wind speeds, and can be quieter, making them suitable for residential areas.

Generally, the higher the turbine, the stronger and more consistent the wind because there are fewer obstacles to air flow. Regardless of how you mount the turbine, you'll want to make sure it's securely installed, and if it's rooftop mounted, properly isolated so you're not hearing a lot of vibration. If it's pole mounted, you'll want to make sure it's not going to damage anything important if freak weather takes down the pole.

Like solar power, wind turbines can be grid-tied, allowing homeowners to sell excess power back to the grid, or off-grid, which involves battery storage for when wind speeds are low.

The cost of residential wind turbines can vary based on size, type, and installation complexities. While the initial investment can be significant, many regions offer tax credits, rebates, and other incentives to offset these costs. And again, don't forget to factor in the potential savings on electricity bills.

Turbines will require a bit more maintenance than solar panels, so you'll need to factor that in too. Regular inspections can identify wear and tear on blades, bearings, and electrical components. Most turbines come with a warranty of 20 to 25 years. Do your homework on available systems and installers here as well, to make sure you get the best value for your buck.

Microhydropower

Microhydropower is a bit of a mouthful, but it's about transforming the energy of flowing water into rotational energy, which is in turn converted into electricity. You're likely used to thinking of hydro power coming from big dams, but itty-bitty systems exist too.

They don't apply to most homeowner situations because we don't all live next to flowing water, and there's not a lot of readily available commercial systems right now anyway. So why mention it?

I bring the idea to your attention because there are some interesting new technologies being worked on that might be available either for residents or at the municipal power level in the next few years.

One is a whirlpool turbine. Touted as being fast to install and fish-friendly, it's an installation that takes advantage of a smallish change in water level and a circular concrete channel that creates, you guessed it, a mini-whirlpool. It's enough to produce from 15 to 70 kW of constant energy, which would power dozens of homes.

Scientists have also been working on ways to generate power from raindrops. It sounds weird but, in the lab, a drop of just 100 microlitres of water released from a height of 15 centimetres can generate a voltage of over 140V. The power generated can light up 100 small LED lights. The goal will be to see how we can scale that up.

And folks who live in rainy communities, or that have a whole rainy season, have rigged up mini-rain gutter turbines to take advantage of all that sluicing water.

As you can see by the range of ideas, some are currently more practical and efficient than others. The point here is to get you thinking in new ways about energy production, how we might do so in all weather conditions in the future, and how we might convert mundane objects like windows and downspouts into power generators.

Lights

One other significant, yet often overlooked issue demands our attention: light pollution. This section delves into the definition of light pollution, explores its implications for wildlife, and offers some guidance on how to mitigate it.

You're Kidding, Right?

I'm not. It's a thing!

Light pollution is excessive and/or misdirected artificial light. It's a byproduct of industrial civilization and a hallmark of urban sprawl. If you live in a city, you probably don't even notice it, but if you take a trip out to the countryside, far enough from an urban centre and look up on a cloudless night, you'll be able to see the difference. In the city, you'll be lucky to see the moon and a handful of stars. Out in the country, you'll be able to see the hundreds of thousands that the city glare washes out.

The Effects on Wildlife

The impacts of light pollution extend far beyond the aesthetic loss of starry nights though. Your local flora and fauna rely on the rhythms of natural light. For billions of years, life on Earth has evolved under a cycle of day and night, with organisms developing behaviours tied to these environmental cues. The introduction of artificial light disrupts these patterns, with profound consequences:

Nocturnal Animals: Species such as bats, owls, and nightjars, which are adapted to hunt under the cover of darkness, find their activities hampered by artificial lighting. The altered conditions favour their prey, leading to imbalanced ecosystems and declining predator populations.

Migration Patterns: Many migratory birds navigate by the stars. Light pollution obscures these celestial guides, leading to disorientation and fatal collisions with illuminated buildings and towers.

Reproductive Cycles: For creatures like fireflies and turtles, light plays a crucial role in mating behaviours and nesting. Artificial lights can deter fireflies from flashing, disrupt turtles' nesting activities, and mislead hatchlings away from the safety of the water, increasing mortality rates.

Plant Life: Plants, too, are affected by artificial lighting, which can alter growth patterns, flowering times, and leaf drop, potentially disrupting the interconnected web of pollinators and seed dispersers.

Bug Death: Many insects, including important pollinators, are attracted to artificial lights, and either burn up on the hot surface of the bulb, or waste their lives mindlessly crashing into the light.

Mitigating Light Pollution

We're all a little afraid of the dark, and we need to balance the need to live our lives safely with our environment. Here are several steps homeowners can take to contribute to a darker, more natural night:

Use Outdoor Lighting Wisely: Install lights only where they are needed, ensuring they are directed downward to minimize skyglow. There's a lot of what amounts to vanity lighting on houses these days, and while we can appreciate people taking pride in their home, it doesn't have to have the full magazine cover level light show on every night. Consider motion sensors and timers to reduce the duration of illumination.

Shield Outdoor Lights: Ensure that all outdoor fixtures are also shielded, casting light downwards rather than sideways or upwards. This simple adjustment can dramatically reduce light trespass and skyglow.

Choose Appropriate Bulbs: Opt for warm-coloured, low-luminance LED bulbs. These are not only energy-efficient but also less disruptive to wildlife, as they emit light closer to the red spectrum, which is less attractive to insects and less likely to interfere with nocturnal animals.

Advocate for Dark Sky Initiatives: Support local and national efforts to adopt dark sky policies, including the use of dark sky compliant lighting in public spaces and the establishment of dark sky preserves.

Educate Your Community: Share your knowledge about light pollution and its impacts with neighbours, schools, and local officials. Collective action can lead to community-wide improvements in lighting practices.

Embrace the Night: Finally, cultivate an appreciation for the natural night sky. Spend time stargazing and encourage your community to host dark sky events. A cultural shift towards valuing the night can be a powerful force for change.

Travel and Commuting

Travel

Traveling to see loved ones or to take a break is common, but it can contribute a lot to your carbon footprint. Here's a few ways to cut down on the impact while we wait for our transit systems to become greener:

1. Fly less often. Flights are big contributors to emissions, and it will be a while before we have reliable electric planes. In the meantime, cut back on flying. This is one area where we're going to have to cut back in the short term.
2. Take the train instead. Instead of those cheap short-hop flights (which, because of their more frequent takeoffs and landings are more carbon intensive), take a train. Less time waiting in the airport, better food, and drink too.
3. Avoid cruises. Sorry, I'm a wet blanket on this one too. Cruises, as they're done right now, are

absolutely terrible for the environment. A large ship can use as much as 250 tons[1] of fuel in a single day. One cruise ship produces roughly the same amount[2] of carbon emissions as 12,000 cars! The air quality on ships is also bad because they burn poor quality, cheap bunker fuel, which puts out lots of black carbon, sulphates, and other chemicals. Some companies have been caught dumping trash overboard, but it's legal for them to dump untreated sewage into the ocean as long as they're far enough offshore. Ew!

4. When in Rome... or any other place, take public transportation. Or use local tour buses. Public transport reduces emissions per person.
5. If you must fly, take a direct flight (less intensive), and pack light. The more a plane weighs, the more fuel it uses.
6. While on holiday, support local economies. The snow globe in Rome is made in the same factory in China as the snow globe in Paris. Skip the tat and go for something nice and locally made.
7. Reduce, reuse, recycle on the road. Avoid single-use plastics, carry reusable water bottles and bags, and recycle whenever possible.
8. Conserve water and energy. Same as you would at home, be mindful of your water and energy use in hotels. Simple actions like turning off lights and AC when not in the room, taking shorter showers, and reusing towels can make a significant difference.

9. Travel during off-peak seasons. Traveling during the off-peak season can reduce the strain on popular destinations, helping to preserve them and reduce the overall environmental impact. It also often results in a more enjoyable and personal experience, avoiding the crowds of peak travel times. Obviously, you'll want to make sure your destinations will still be open, so plan ahead.

The Dreaded Commute

We've normalized the commute and it's not good for anyone. Wear and tear on your vehicle, the insurance costs, and worse, the hours of your life eaten up staring at the back of someone else's car. And unless you're very lucky, you don't get paid for that time. Ugh. What can you do?

1. Demand work at home time. One of the few silver linings of the pandemic was that plenty of those meetings could, in fact, be emails. Companies managed to conduct business just fine with everyone at home. Push to keep that or get it back.
2. Consider moving closer. Obviously, many factors play into where you live and work, especially affordability. But have you run the numbers on how much time and money it's taking you to get back and forth? Have you done so recently? (Considering that expenses have been outpacing

wage raises for decades now, it's worth looking at again.)

3. Change your mode of transportation. Is it possible to carpool? Are you able to switch to a hybrid or an EV? Could you walk, bike, or take public transit? Everyone is different of course, but personally, I'd rather sit on a train and either get some work done or relax with a book than spend time behind the wheel.
4. If you're stuck with the current situation, do at least make sure you're using the most fuel-efficient route, and skip those drive throughs on the way to and from work.

Inside Your Home

Bathroom

No one likes to talk about bathrooms much, but given how important they are to our lives and health, it's a great place to start our review of how to make our homes more sustainable. So, let's talk about...

Toilets

These little porcelain thrones are a tremendous source of waste in all senses of the word. There's a lot we can do to make them better.

If you wish to use less water per flush, sticking half a brick or a filled water bottle in the tank reduces how much it fills up (this is called tank displacement), which then means you use less water per flush. You may have to experiment a bit to get the optimal level of displacement that is still effective.

If you have to replace your toilet for whatever reason, you can install a dual flush toilet. As the name implies, these have two options, one for liquid waste and a higher-volume flush for solid waste.

As for toilet paper, you can choose to source yours from recycled materials or more sustainable materials like hemp or bamboo, so we're not clear-cutting forests just to wipe bums. You could also retrofit your toilet cheaply with a seat-mounted bidet, a device that sprays water at your posterior to clean it. You can then dry with a reusable/washable cloth and skip the toilet paper. If you share a bathroom, you might use colour-coded cloths.

Give the pre-packaged wet wipes a miss. Despite package claims, they're not *really* flushable (sewage maintenance staff around the world hate them), don't break down easily, use a lot of material, and are heavy to ship because of the moisture content.

While we're on the subject of what's flushable, remember your toilet is not a trash can! At most, only two things should be going down that pipe, toilet paper, and your, ahem, biological emissions. Dead critters, candy wrappers, cigarette butts, dental floss, bandages... all these things should be disposed of elsewhere.

Speaking of bathroom paper products, facial tissue can be replaced with reusable handkerchiefs, or at the very least, made with recycled materials.

If you do end up flushing too much down the toilet and end up with a clog, a reusable "mechanical snake" or "plumber's snake" is a great tool to have on standby. Basically, it's a long, stiff metal rope you can push down the toilet and twiddle around to deal with a backup. Be sure to clean it off after use.

If you *really* want to go the extra mile, you can install a grey water system to use sink and shower water for flushing the toilet.

Diapers

Diapers are terrible for the environment. The sheer volume of used nappies a single child produces is staggering: anywhere between 2500-4000! They're also nasty and smelly, so an entire industry has sprung up for 'diaper management.' Of course, this involves a LOT of plastic: plastic bags, bins, plastic packaging...

I get it though: new parents are very tired, barely able to keep up with the demands of life plus children, and disposables are very convenient. Cloth diapers look like a lot of work. Yucky work, to boot.

However, there is a happy medium: cloth diapers with flushable liners. The liners are a somewhat stronger toilet paper, and thus solid waste is easily removed from the diaper and put down the toilet. The reusable diaper can then be put in a bucket with a lid, and when full, the entire lot can be thrown in the washing machine.

Baths and Showers

Let's get this out of the way first: you *might* be washing more than you need to.

Don't get me wrong, personal hygiene is important, and body odour is not nice. But let's be aware of how much *marketing* pressure we've been put under in the Western world when it comes to cleanliness. Chances are, if you work a desk job in a climate-controlled building, you're not getting dirty or sweaty enough to really *need* a daily cleaning. So, consider skipping one now and then.

As for baths vs. showers, I like a hot bath as much as the next person, but since they're so water intensive, I save these as a treat. In the shower, we've installed a low-flow shower head with multiple settings; the pressure remains fabulous and there's a lovely mist setting that's like being wrapped in a warm foggy hug. Obviously, shorter showers are less resource intensive than longer showers, both in terms of water heating energy and water usage.

If you can afford it, and especially if you have young kids that will be in your household for years to come yet, get an instant hot water heater, and/or supplement your supply with a solar water heater. They'll reduce your energy bills and ensure that everyone has access to hot water even on higher usage days.

Oh, and hey, bath bombs, salts, and oils? A lovely indulgence. They get rinsed down the drain and into the water supply though, so make sure they use biodegradable ingredients, and come in recyclable packaging.

Toiletries

The next biggest source of waste in the bathroom are the products we use to keep clean. There are SO many products, and we've normalized way too many "disposables." Once again, it's important to be conscious of what we *need* versus what we've been told — over and over and over again — to want. We also need to be really conscious of packaging waste here.

This is a long list, and you're not going to be able to work your way through it in one shopping cycle. Start at the top and work your way through them as you use up what you already have.

Toothpaste. Definitely a necessity, but probably one of the worst offenders, packaging wise. The tubes are not recyclable in most jurisdictions and so much toothpaste gets left behind even with the most dedicated squishing methods. Toothpaste tablets and powders are now more accessible, so shop around for those. Do be careful though: some come with xylitol as an ingredient which is not safe around pets, dogs especially. One other thing to avoid is any toothpaste that has "anti-bacterial action" in the form of something like triclosan. It's more than you need, and we're rinsing too much of this stuff into the water supply. There's also evidence to suggest that overuse of antibiotics is causing the bacteria to evolve to resist such things, which becomes dangerous when you need to fight a life-threatening infection.

Mouthwash. Here too, there are strips and tablets, dissolvable in the mouth or in water, that get the job done with less packaging.

Toothbrushes. We touched on these earlier in the book: millions of these go to dumps every year. Bamboo brushes are at least biodegradable and so a reasonable alternative in the short term; the bristles can be removed when they age out. There are also toothbrushes with replaceable heads, so that you're only replacing a smaller portion.

Shampoo, hand soap, and conditioner. There are a couple of ways to be more sustainable with these kinds of products. With liquid versions, you can save a lot of packaging and shipping by buying refillable dispensers and buying in bulk. On the other hand, there is the argument that liquid versions still use up more packaging overall, and are heavier to ship, incurring a larger carbon footprint, and are more water intensive. Bar versions might be better if you can keep them dry between uses and find a way to combine all those small, hard-to-handle slivers of soap rather than binning them. The choice here will come down to what produces the least amount of waste and emissions at *your* house.

Body wash and plastic scrubbies. There's some evidence to suggest that using a lot of soap too often can dry the skin; of course, corporations are only too glad to sell us moisturizers to fix this problem! You're the best judge of your own body, of course, but consider whether one habit is creating a problem that doesn't need to exist. As for plastic scrubbies, exfoliation might be necessary from time to time, but buy biodegradable options like loofahs. The scrubbies will be contributing to microplastic waste, and aren't recyclable, nor can they be repurposed easily.

Perfume. To me, most of what is marketed as fragrance is overpowering, and gives me a splitting headache. I'm not alone in that, as many places now insist on being fragrance free. Strangely, there are no regulations forcing manufacturers to list perfume or fragrance ingredients, but we know that many synthetic scents contain phthalates, which interfere with sex hormone production, among many other health impacts. I would encourage you to go fragrance free both in terms of the perfume bottle and the scents added to all our toiletries. If you must have a scent, look for 100% "organic" or "natural" options (but beware there aren't regulations that govern what can be claimed by those terms.)

Shaving cream, razors, beard oils. What you shave and how often is an entirely personal choice, but it too is highly influenced by marketing and especially social media. For a while there, big hipster beards were all the rage, and now we're back to shaving again. Changing styles frequently, you see, sells more products! The worst product in this category is the disposable razor. They're cheap, not even recyclable, and don't provide a great shave anyway. The razors with replaceable heads are less wasteful, and if you can find a place to recycle the heads, that's great. Better are the new generation of straight razors where all you need to replace is the blade itself;

they're cheaper in the long run as well. If you shave your face, you can replace those hard-to-recycle pressurized cans of shaving cream or gel with old school shaving soap. If you shave your legs and underarms, a surprising number of people just use their hair conditioner as a lubricant, as it's cheaper and just as effective as anything else, especially if you can buy it in bulk.

Deodorant and anti-perspirant. If you can get away with just a deodorant, you can shop around for one that comes in sustainable packaging or can be refilled, as there are now dozens of choices in this category. As of this writing, there don't seem to be nearly as many options on the anti-perspirant side; I was able to find just one roll-on product in the UK. But do search! Product developers use search engine data to judge the market for their ideas, and if they see many people are looking for something, it increases the chance that it will come out.

Swabs/ear buds. Doctors advise against sticking *anything* in your ear, and recommend just wiping the edge of the ear canal with a warm, wet cloth. If you do need to clean your ears, avoid the swabs with plastic stems. Even better from a sustainability perspective are the reusable, washable stainless steel coil tips, but for heaven's sake, be gentle and don't poke too far in with any of it.

Menstrual and incontinence products. Standard pads and liners have long used bleached fibres and plastic backings and individual wrappers. Tampons are notorious for their plastic applicators; they wash up on shore so often as to be called 'beach whistles.' There are now brands of both products that use biodegradable materials, so please, use them instead of their plasticized cousins. You can also consider a menstrual cup, which, with proper washing and care, is reusable.

Condoms. This one is tricky, because as I mentioned in the section on population, birth control choices are critical to a sustainable future. Not to mention that we absolutely need protection against sexually transmitted diseases. But condoms are made from latex or polyurethane which take forever to degrade and are packaged in foil or plastic. The best you can do here is start complaining to manufacturers about packaging. Please don't leave used ones just lying around outside either, dispose of them properly! Not only is that just gross, unfortunately some animals and larger aquatic creatures will eat them, and that messes up their digestive tracts.

Makeup, makeup remover, facial toners, and moisturizer. Makeup packaging is a disaster, frankly. Eyeshadow, for example, comes in plastic trays that aren't recyclable, in amounts that are never calibrated so you use the whole package at once (there's always one shade you don't use as much of as the others!) and come with those useless plastic and foam applicators. Everything else is just as bad: lipstick, mascara, foundation, hand cream... The good news is that there are companies that are addressing this. My own eyeshadow, for example now comes from a company that sells the individual shades in small metal containers that fit in a reusable bamboo box, and that can be applied with a natural fibre bristle brush with a bamboo handle. As for makeup removers, you can use a washcloth, rather than disposable cotton balls, or worse, foam sponges.

Floss. A lot of these are plastic or plastic coated. Look for bamboo or hemp fibre-based flosses in non-plastic containers.

Nail files. Skip the disposable emory boards and invest in a metal file.

The Rest of the Room...

We don't need to buy them very often, but when the time comes to replace bathmats, towels, face cloths, and natural fibres are best. Look for cotton, organic cotton, bamboo, or hemp. Synthetic fibres don't break down easily and a lot of fibres get washed out into the water supply where they choke out wildlife. Reuse the remains of your bathroom linens as rags, for towelling down pets, for camping. At the end of their lives, they should be compostable or recyclable. If you use a shower curtain, avoid PVC versions, and get one that doesn't cling so you're not spilling a lot of water outside the shower and wasting it.

As for our medicine cabinet, we can make better choices there too. Our prescription meds obviously can't be switched up, but you can email the manufacturer and ask them for better, sustainable packaging. We do have choices when it comes to our over the counter (OTC) drugs, and we go through a *lot* of these. When you can, avoid blister packs, or sheets of individually wrapped pills, choosing bottles of pills instead. And again, take a moment to email the manufacturer and ask them to do better.

Kitchen

The kitchen is the centre of your home. It's the modern-day hearth, where we 'break bread.' It's also a major source of waste production. Let's see where we can reduce our impact and save a lot of money to boot.

Reduce Your Use of Disposables

We've all been guilty of turning to disposables for convenience, but these items are major contributors to the global plastic pollution problem and the global waste problem. Worse, we're throwing away money every time we use a disposable. Here's a quick inventory of what you might be using and what you can replace them with:

- Paper towels. These are fine for big spills but should be used sparingly otherwise. Instead, use washable bar towels, wash down cloths, or even cheap face cloths for regular surface cleaning. We have a big pile of these and a little hamper for them in our laundry room to hold the dirty ones until a load of wash goes through. If you do buy paper towel, try to buy paper towel made of recycled paper, and consider disposing of it in the composter if you're wiping up something organic.

- Paper plates and cups. Because these are plastic or wax coated and/or have plastic cores, or worse, made of Styrofoam, these are terrible. They use up a lot of resources, and generally aren't recyclable, even if the packaging says they are, because recycling plants can't deal with whatever might be stuck on them. Stick with the real stuff.
- Plastic wrap. I shudder to think how much single use plastic wrap is just lying about in dumps around the world. And there's no need for it! Use sealable containers instead; either stuff you've saved from the grocery store and put into use or specialized containers with lids. Not only are these easier to handle (how many times have you cursed the plastic wrap for sticking to itself?!) they come in a variety of sizes and can be washed to stay clean. You can also use beeswax wraps if you want something prettier to cover your dishes with.

Conserve Water

Water conservation is key if you want an eco-friendly kitchen. Turn off the faucet while washing dishes and use a bowl while washing fruits and vegetables instead of running them under water from the tap. You can also install low-flow aerators on sink faucets which will help conserve water without compromising efficiency or power. This will save on your water bill, and the water will taste better too.

Washing dishes? Dishwashers are more efficient, because they use high pressure jets and insulated heat to make water use more efficient. Use phosphate-free soap no matter which method you use to wash dishes, and eco-friendly rinse aid for the dishwasher. Remember, all of what you wash or rinse with goes back into our water supply; too many phosphates cause algae blooms, which then choke off all other aquatic life.

Making Your Appliances Greener

The appliances in your kitchen are responsible for a large portion of your energy consumption. This means they have a significant effect on your carbon footprint. To reduce your environmental impact, opt for energy efficient models when replacing or buying new appliances. Look for those with an Energy Star rating which indicates that it is more efficient than other models on the market. Make sure you follow local guidelines for getting rid of old appliances, so they are properly recycled.

Even with an Energy Star rated oven, it will still be the biggest energy hog in your kitchen, so use it wisely. Cover pots of water to bring them to a boil more efficiently, and don't worry about preheating the oven for things like your baked potatoes or bacon. Just stick the things in there as it's warming up.

Greener also requires some cleaner: ovens and fridges run better when the oven isn't full of baked on crud, and the fridge coils (at the back) aren't covered in dust.

... and No More Gas Stoves

While some cooks swear by gas-powered stoves, there's an increasing body of evidence [1]to suggest that they're big contributors to indoor air pollution. Bad installs, or vents that haven't been maintained, or forgetting to turn on the hood extraction fan when you cook: all of these can mean you may have small methane leaks, increased nitrogen dioxide, and trace quantities of benzene and other volatile organic compounds in your house. These can cause respiratory and other issues. And of course, methane (natural gas) is a powerful greenhouse gas. The less we use it, the less we're likely to release it into the atmosphere. We use an ordinary electric range; a lot of professional chefs are now discovering that induction cooktops are fast and even. There's no need to junk your current stove if you use gas right now but do get the installation inspected and cleaned up if necessary and make sure you're venting properly. When it comes time to replace, consider that induction stove.

Instant Hot Water Heater

Also known as tankless hot water heaters, these are much more efficient because they heat water as you need it, rather than keeping a large tank of water warm all the time. The upfront costs of these are still on the high side; you might want to see if there's a government subsidy available to make the switch. The operational cost savings are great, though, and you don't ever "run out" of hot water either. Supplementing with a solar water heater might be an option too.

Reducing Food Waste

Food waste is another major contributor to our carbon footprints, so reducing food waste should be a priority for anyone looking to make their kitchen eco-friendlier. Start by being mindful about how much food you buy and use — only buy what you need and try to use up all leftovers before they go bad!

I have a section on compost elsewhere, but briefly: composting food scraps is another great way to reduce waste — this helps break down organic material into nutrient-rich soil without releasing as much of the harmful greenhouse gases into the atmosphere that landfills do.

Opt For Sustainable Materials

Another way to make your kitchen more eco-friendly is by using sustainable materials when possible. Consider, for example, coconut "coir" or loofah scrubbies for dish cleaning, and bamboo handled utensils instead of plastic.

Cookware

Non-stick pans are convenient but there's evidence to suggest that they release toxic chemicals into the air and your food when heated. Look for eco-friendly cookware that is free of toxic chemicals like PFOAs and non-toxic materials like cast iron, stainless steel, and ceramic.

Cleaning Products

I'll mention them here because a lot of us store them under the kitchen sink. Use eco-friendly cleaning products: Standard cleaning products often contain chemicals that can damage the environment. Choose eco-friendly alternatives such as baking soda, vinegar, or natural soap instead. If you don't have the time or patience to make and bottle your own eco-friendly cleaners (I don't!), there are an increasing number of these available at the grocery store. Even better, many of these come in large refill sizes.

Use rags, old towels and facecloths for actual scrubbing and cleaning, and instead of disposable dust mops and wash mops, get washable, reusable heads.

Packaging, Packaging, Packaging

I saved this for last, not because it's the least problematic issue in the kitchen. In fact, it's the worst! But it will take a bit of sustained effort on your part to whittle this down.

So much of what we bring home from the grocery store involves plastic these days. It's disheartening. I've seen oranges that have been peeled and then put into plastic containers. I've seen cucumbers that are shrink-wrapped. I recently opened a box of spaghetti — a dry good if there ever was one — and discovered an additional plastic bag enclosing the pasta. Argh!

All in the name of convenience, supposedly, although I don't recall any customer petition asking for pre-peeled oranges. It's one of those things marketed to us to sell more oranges... at the expense of the environment.

What can you do? Pick a product you regularly buy and look at it more critically. Going back to pasta, for example, it typically comes in either a single use plastic bag, or a cardboard box. The cardboard box is at least recyclable. The plastic, not so much. Find a comparable brand that uses better packaging.

Tea bags... a lot of them now come in plasticized individual pouches, which is unnecessary. Once you open the box of tea bags, all you need to do is put them in an airtight tin to keep them fresh! So, pick another brand of tea that does better. (Also avoid those brands that have plastic mesh tea bags!)

And so on, and so on. You get the idea. If you want to take it one step further, send a polite, but firm note to the company's customer service centre and complain about their wasteful packaging. Tell them you're not buying their product anymore and why.

Another way to reduce grocery store packaging is to buy in bulk and buy refillable products. At the moment a lot of this stuff is only available at higher end or boutique style grocery shops. But some of the bigger chains are getting into this, so use it when you can.

Now we move onto the main point of the kitchen, which deserves a complete section on its own.

Food

The food we eat and the way we produce it are deeply intertwined with the health of our planet. From the fields and farms to our forks, every step in the food supply chain has a profound impact on our environment. This chapter explores practical strategies for adopting a more sustainable diet, focusing on mindful consumption practices that support both the earth and its inhabitants.

Reducing Meat and Dairy Consumption

One of the most impactful changes an individual can make to their diet in the name of sustainability is to eat less meat and dairy.

I understand why people resist this one. Meat and cheese are delicious, very filling, and generally seem easier and faster to prepare.

But livestock farming is a significant source of greenhouse gases, contributing to climate change, deforestation, and water scarcity. By reducing our meat and dairy intake, we can decrease the demand for these resource-intensive products.

If you can't or don't want to be vegetarian or vegan, then consider just committing to plant-based meals more frequently. "Meatless Mondays" are an easy way to cut 15% (one day out of seven) of your meat and dairy consumption without a lot of overthinking. If all of us big meat and dairy eaters cut back by that much, it would make a huge difference.

If you're pressed for time, frozen vegetables come nearly fully prepared (they're cleaned, peeled, and cut up) and are easy to cook with. Pasta is another easy option, as are tinned legumes like chickpeas or kidney beans.

When you do eat meat, save the beef for special occasions. Not only is it super expensive, but it also has the biggest environmental impact. The chart on this page[1] from *Our World in Data* gives the current numbers for worst to best protein sources from an environmental point of view.

Alternative "milks" are a more complicated question. Some products are loaded with sugar to make them more palatable, which isn't great from a health standpoint, others are super water intensive to grow. Almond milk, for example, is tasty but requires thousands of litres of fresh water per tree and they're primarily grown in areas that require irrigation. Cashew milk, meanwhile, is associated with exploitative labour practices, and the rush to provide enough coconut milk to supply the market has led to further habitat destruction in tropical countries where farmers are desperate for income. And of course, your choice will also be constrained by food allergies and what you're using the product for. So, for this one, take the time to see what products are available to you, investigate how they're made, and switch between two or three.

Eating Local

Sourcing food locally reduces the carbon emissions associated with long-distance transportation and supports local economies. Seasonal eating not only ensures that you're enjoying produce at its peak of freshness and nutritional value but also connects you more deeply with the natural cycles of your region. Farmers' markets, community-supported agriculture (CSA) programs, and even home gardening are excellent ways to embrace local foods.

Speaking of home gardening, if you have the space and the time, I recommend trying it at least once. It likely won't save you any money, especially when you include your labour, but it is very satisfying to bring super fresh produce to your table. It will also give you a much greater appreciation of how challenging it can be to fend off insects, animals, and the weather long enough to produce something edible.

Cooking at Home

Yes, this is a way to be more sustainable! Preparing meals at home offers more control over the ingredients used and reduces food waste and packaging, especially when compared to fast food.

If you're *really* organized, you can cook in batches, freeze additional portions, and save time and money. Personally, I'm not there yet, but this might be for you. Or even a few of you in the family working all at once! Many of the things we buy premade at the store these days – like perogies, for instance – were traditionally made by groups of people working all day together, assembly line style. You could apply the same approach to batch meal creation.

Choosing Fair Trade

We touched on this before, but it's worth bringing up again in this section. When buying products such as coffee, chocolate, and bananas, opting for fair trade certified options ensures that producers in developing countries receive a fair price for their goods. Fair trade practices promote sustainable farming methods, improve working conditions, and support community development projects. By making conscious choices to support fair trade, you can contribute to a more equitable global food system.

Avoiding Seafood or Choosing Sustainable Options

With overfishing posing a significant threat to marine ecosystems, reducing seafood consumption, or selecting sustainable options is vital.

For those who choose to eat seafood, it's important to select species that are not overfished and to opt for catch methods that minimize bycatch (where they scoop up *everything* and then chuck away what they're not selling) and habitat destruction. Organizations like the Marine Stewardship Council (MSC) provide certifications and guides to help consumers make informed seafood choices.

Farmed seafood and fish are also options, but again, one must look at how it's done. Ripping up existing ecosystems to plonk down an industrial fish farm is not what any of us have in mind for sustainable food. Check out producers like GreenWave, who are trying to take a regenerative approach.

Flour Power

You can help farmers by cooking with alternative flours. Spelt or wheat can be grown in rotation with rye and clover to replace lost nitrogen in fields without the use of synthetic fertilizers. By baking with these alternatives you provide income to farmers and help them reduce their costs at the same time.

Heating & Cooling

A big source of emissions and yes, pollution are our home HVAC systems. That acronym refers to "heating, ventilation, and air conditioning." They're also a significant source of financial pain, and maybe even marital strife! (Every home seems to have someone turning down the thermostat, while someone else is always turning it up.)

There's a lot we can do to improve our systems, saving money first in the short term, and then when the time comes, upgrade or replace the main units, investing in better, more efficient technologies that also reduce emissions. It's a win-win.

If you rent, you can either skip this chapter or, if your landlord is approachable, have a talk with them about what they can do to improve your building. As always, remember to present it in terms of what's in it for them: reduced costs. Push them to investigate financial incentives from the government. If they're not approachable and are in fact in violation of some local laws, perhaps a quiet (anonymous) word to the authorities to get them to clean up their act.

But first, let's understand how they contribute to the environmental problems we're facing.

Poor Energy Efficiency

You likely "inherited" your existing HVAC system when you bought your residence. It might not be optimized for energy efficiency, leading to excessive energy use. That's not good to begin with. But this increased demand probably relies on fossil fuels, either directly (e.g., a gas furnace) or indirectly (electricity pulled from a fossil fuel powered grid). Fossil fuels, when burned, emit a lot of things, including carbon dioxide (CO^2). CO^2 as we know, is a leading greenhouse gas contributing to the warming of our planet. Methane is *also* a greenhouse gas, and in fact it's worse than CO^2 for trapping heat in our atmosphere. So being able to reduce our use is important.

Those Refrigerants

If your AC unit is old, it still might contain a coolant type called a chlorofluorocarbon, or CFCs. Remember the news a long while back about how we were thinning the ozone layer, that protective layer of our atmosphere? (It's essential for protecting life from the Sun's harmful ultraviolet radiation.) CFCs were the culprit. We banned them (check out the Montreal Protocol), and fortunately the ozone layer is returning to normal.

Unfortunately, what we replaced them with — hydrofluorocarbons, or HFCs — are themselves a problem because they too trap heat in the atmosphere. HFCs leaking from cooling equipment are thought to contribute about 4% of global greenhouse gas emissions[1]_— about twice as much as aviation. So, we're having to update our coolants again with lessons learned from both problems.

Waste

Did your parents (or maybe your life partner, see marital strife, above) ever use the phrase: I'm not paying to heat the outdoors! Well, I hate to say it, but... you're probably paying to heat the outdoors. Or cool it, depending on the season.

Houses are notoriously leaky when it comes to heating and cooling, either through actual drafts or through poor insulation. More waste means you're paying more to keep the house at a constant temperature. And of course, using more energy.

Let's see how we can fix these issues.

Level One - Tune Up the Existing System

Enhancing the efficiency of your HVAC system and home insulation doesn't require a hefty investment. Even with a modest budget, there are several practical steps you can take to improve your heating, cooling, and insulation, ultimately reducing both your carbon footprint and energy bills. Let's do heating first.

Forced-air gas furnace

If you have a forced-air gas furnace system, one of the simplest and most cost-effective steps is to clean or replace your HVAC filters every few months. Dirty filters restrict airflow and reduce system efficiency.

Inspect your ductwork for leaks or separations and use mastic tape or sealant to repair them. This prevents air loss and improves the efficiency of your system.

If you have a little more money to spare, have a trustworthy HVAC system company inspect and adjust your system. Pro tip: do this in the off season! We had a furnace fail in February and when the company came in to fix it... it had to condemn the furnace because faulty gas appliances are dangerous. Of course, at that time of year, it's hard to get a replacement and the house got cold. Get those inspections done in spring or fall where the consequences won't be dire.

Radiant heating systems

Radiant heating systems, such as baseboard heaters or radiant floor heating, provide heat directly to the floor or to panels in the wall or ceiling. These are low maintenance, so if you have one, lucky you! Also, please know I envy your toasty floors.

You should have someone out once a year to clean the boiler, check the flame, flush the boiler, and check the glycol, temperature, and pressure. Do this in the off season, so that if issues are discovered, you won't freeze in the event of a system shutdown.

Hydronic heating systems

Hydronic systems circulate hot water through radiators or underfloor piping. Radiators should be "bled" once a year. That is, you open the little valve on each of these and release some water to get rid of air bubbles. Do this before you turn the system on, as the bubbles may cause hot water to spatter (ouch!). Air trapped in radiators can reduce efficiency. Bleeding them annually improves heat distribution.

You can also insulate the pipes. This reduces heat loss as water travels from the boiler to the radiators.

If you have a little more cash, you can install "thermostatic radiator valves." These allow you to control the temperature in individual rooms, so you can reduce heating expenses for rooms that only need a minimum of heating most of the time.

Electric heaters

Electric heaters are common in homes without central heating systems, or as supplementary heating. Your best bet here is to keep them clean of dust and debris and make use of individual thermostat settings to heat only as much as you need.

Wood stoves and fireplaces

If you're burning wood, be aware that it's not the most environmentally sound option, both in terms of what it outputs and the fact that you're using up trees. At least make sure it's dry and seasoned wood. This produces more heat and less creosote buildup, improving efficiency. You can also get a heat exchanger to move more heat from the stove or fireplace into the room. Finally, make sure your chimneys are clean! This improves airflow and cuts your risk of fire.

That thermostat

This one you know already: Lower your thermostat by a few degrees in winter and increase it in summer. Sweaters and shorts save money! A programmable thermostat, which is relatively inexpensive, can automate this process, ensuring optimal operation.

Now, let's talk cooling

Improving the efficiency of your home's cooling system doesn't necessarily mean investing in a new, high-end air conditioner. There are several budget-friendly strategies to enhance the performance of your existing system, ensuring it runs more efficiently and economically.

Regular maintenance of air conditioning units

Dirty filters restrict airflow and reduce efficiency. Cleaning or replacing your AC filters can improve performance. Also make sure the outside unit mesh doesn't get covered with dirt, debris, leaves and so on.

Where we live, there's a tree that releases a lot of 'fluff' that looks like feathers or dandelion seeds. Too much of that really clogs things up, so I check the AC weekly.

Start a fan club

Use fans to circulate air in occupied rooms. This can allow you to raise the thermostat setting by a few degrees without reducing comfort. Ceiling fans are your best bet, as they're quiet, and can be reversed depending on the season. Otherwise find the quietest, most efficient fans you can.

But also, don't forget about exhaust fans! Use kitchen and bathroom exhaust fans to remove hot air and excess moisture after cooking or showering. This will also help improve indoor air quality.

Manage indoor heat

If you can, avoid heat *generation*. Hang the laundry to dry, use a microwave or air fryer to cook, or do meals that don't require cooking at all. If you still have those incandescent bulbs, turn them off or better yet, replace them with cool LEDs.

Throw some shade

Plant trees or install awnings to shade your home from direct sunlight, especially on windows facing west and south. Close curtains on the west and south sides of your house as well, to avoid what they call "solar gain" during the day. (The opposite is true of the winter of course!).

Venting and timing

If it gets cool enough at night, and it's safe to do so, turn off your cooling system and open your windows. Close them in the morning to trap the cool air.

Otherwise, run the air conditioning a little cooler at night when it has to work less hard to bring down the temperature in the first place.

Annual servicing

Have a professional check your cooling system annually. They can identify and fix minor problems before they lead to bigger, more costly issues.

Level Two - Improving Home Insulation on a Budget

When it comes to insulating your home, you might be worried that it is a costly and complex process, involving tearing down walls or professional interventions. Sometimes it can be, however, there are numerous ways to improve your home's insulation, even with limited resources.

Let's start with the basics: understanding where most heat loss occurs in a typical home. It's often through windows, doors, and especially the attic. Heat rises!

Windows are notorious for heat loss. One economical, although somewhat waste-generating solution is to use insulating window films, a DIY-friendly project that can be done in a weekend. It's a little like sealing your window with cling wrap. These films add an extra layer of insulation and are inexpensive.

Better, but more expensive: you can also use heavy, thermal-lined curtains. Not only do they add a cozy aesthetic to your rooms, but they also trap heat inside, reducing the need for constant heating. In the summer, these curtains can block the sun's heat, keeping rooms cooler.

Doors, especially older ones, can have gaps that let in drafts. Weather stripping is a cost-effective way to seal these gaps. You can find weather stripping materials at any hardware store and applying them is straightforward — it's just a matter of cutting to size and sticking them in place. For the bottom of the doors, consider a simple draft stopper. You can make one use old fabrics or clothing, but you can also get these at the store too.

Attics play a crucial role in home insulation because if your attic is poorly insulated, heat will escape. Adding insulation to your attic can be more of an investment in terms of time and effort, but it pays off. Fiberglass batts (they look like thick fluffy blankets), for example, are affordable and can be installed without professional help. They fit between the beams of your attic floor. Tread carefully in an attic! Stick to the beams and don't put your foot through the ceiling. I'd recommend wearing coveralls when installing these, and definitely use gloves and a mask, as insulation materials are not good to breathe and fibreglass splinters suck.

Let's not forget about the walls. While adding insulation to walls can be more intrusive and expensive, it's worth it, especially for exterior walls. It might be possible to pour a loose fill type of insulation down the inside of the walls from the attic, or maybe you use insulating panels on top of existing walls. Consult a professional for this type of job, however.

Lastly, consider the power of rugs and carpets. Not only do they add warmth and style to your home, but they also provide additional insulation to your floors, especially if you have hardwood or tile flooring. They cut the echo in a large room, too! They also manufacture washable rugs these days, making them easier to care for.

Level Three - Upgrading the HVAC System

If your existing system is up for replacement, now would be a good time to change things up. At the very least, you should try to go for the highest efficiency unit you can afford, but if you can, try to get off fossil fuels altogether. Here's some alternatives:

Geothermal heat pumps: Harnessing the Earth's constant underground temperature, geothermal heat pumps offer an energy-efficient solution for both heating and cooling. These systems transfer heat between your home and the ground or a nearby water source. According to the U.S. Department of Energy, geothermal heat pumps can reduce energy consumption by up to 60% compared to traditional systems.

Solar HVAC Systems: Solar-powered HVAC systems use solar panels to collect energy from the sun, converting it into power to run heating and cooling systems. They can reduce reliance on traditional energy sources, leading to lower utility bills and a smaller carbon footprint. If you're worried about the number of sunny days, you can also supplement with battery storage, and rooftop wind turbines. A lot of research is being done on making turbines smaller, more efficient and there are some nifty designs out there, including some that are bladeless!

Air Source Heat Pumps: These absorb heat from the outside air to heat your home and can also function in reverse for cooling. Newer models are effective even in colder climates and can reduce electricity use for heating by approximately 50% compared to electric resistance heating like furnaces and baseboard heaters.

Hybrid Heating Systems: These systems combine a gas furnace with an electric heat pump, allowing flexibility and energy efficiency. They automatically switch between burning fossil fuels and using electricity based on temperature thresholds, ensuring efficient energy use. These would be good systems in climates where it routinely gets down to -30°C. Brr.

Show Me the Money

Governments around the world have emissions targets to meet, and that means they're prepared to hand you some money to help get the job done.

You might find you are eligible for anything from a free "energy audit" of your home, to rebates, to straight up grants. Start at the federal level for incentives and tax credits, but also check your state/provincial/regional and municipal governments for programs as well. And your local utility! You might be pleasantly surprised at what's available.

And if nothing is available, well, time to call your representative and ask why!

Closets and Laundry Rooms

Clothes are a vital necessity, but they're also a hazard. At least, they are the way we do things right now.

Making clothing requires enormous amounts of water. It's hard to find an accurate figure on exactly how much goes into the oft-cited 'single pair of jeans,' but the consensus seems to be more than a thousand litres. (I suspect the numbers vary so widely because of different grades of denim, different dye processes, accuracy of accounting etc.) Even if it's not that high per pair, let's say it's only 10 litres per pair, the sheer number of jeans we churn out annually means water usage is sky high.

As is water pollution. Those dyeing processes can be toxic: the discharge can contain carcinogenic chemicals, salts, and heavy metals. Unscrupulous factories often dump the stuff into the drain or the waterways, untreated.

Energy use is also an issue. There's the energy to get the raw materials from the farms (or in the case of synthetic fibres, factories) to the textile factory, from there to where it's cut and stitched, and then onward to the store. Child labour is still a major problem in a lot of developing countries where clothes are made.

At the store, anything that doesn't sell often gets dumped, tags and all, out back of the mall. And a *lot* gets trashed, because manufacturers make these things in the hundreds of thousands, basically on spec. Too many in a shade no one likes this year? Into the trash. Too many in size three? Into the trash. Quite often deliberately slashed or otherwise ruined[1].

Some of the dead inventory does get sold on to liquidators for pennies on the dollar, where it ends up in dollar stores or more upscale resellers for the big brand names.

What makes it into our closets, either direct or via liquidation, doesn't fare much better though. There's a multi-billion-dollar industry whose sole purpose it is to sell you more clothes, so they set trends accordingly. One year skirt hems will be mid-calf, next year, that will be 'out of date' and only hems above the knee will do. Colours are fashionable one season and laughable the next. It's one of the biggest sets of stories we tell ourselves.

This leads you and me to always be updating our wardrobe, or worse, making impulse purchases of clothes we wear once and then shove to the back of the closet. Eventually we clean it all up and either trash it or send it to a thrift store.

Unfortunately, thrift stores are drowning in our cast offs, so a lot of *that* gets baled up and shipped to developing countries. Who also can't keep up[2].

And because of the law of big numbers, even the simple act of washing your clothes has a huge impact, because there are so many of us doing it.

Sadly, as much as I've tried to escape it, laundry remains an inescapable part of life. So does the requirement to wear pants, at least in public.

So, what can we do to make clothing and laundry lower impact?

Clothes Should Be Thought of as an Investment

Seriously, we need to stop 'fast fashion.' I totally understand the desire to look good and the need for novelty, but it's gotten way out of hand, as noted above. If you can't stop buying new clothes right away, at least cut back to one or two pieces per season. Take your time, consider how it works with everything else in your closet and build what they call a 'capsule wardrobe.' This is where you have a variety of versatile pieces that can be mixed and matched to achieve a variety of different outfits. Consider it a fashion challenge! How low can you go, yet still have lots of different looks?

When clothes do wear out, repurpose, reuse, and recycle at home, or if you're as hopeless with a sewing machine as I am, get someone crafty to help you out in exchange for something they value. Don't count on the thrift stores to handle your cast offs, there's just too much out there.

Synthetic Fabrics?

Let's also consider the clothes themselves. Synthetic fabrics — think polyester, nylon etc. — have many positives, but they're not particularly earth friendly. They're hard to recycle, so often end up in the dump, where they take decades to break down, if at all. Washing plastic-infused clothes releases micro-plastics into our waterways, where they accumulate in ocean wildlife. As you replace your clothes, consider using natural fibres.

What, Like Cotton? Isn't it Bad?

It's true that cotton processing, as noted above, is a water-intensive affair. We should also acknowledge that cotton and the associated textile industries also have a horrendous socio-political history with impacts that reverberate today. The slave trade, industrialization, child labour, worker exploitation, and more broadly, colonization are just some of the issues intimately tied to the uses of this plant. These topics are huge, and beyond the scope of this book, but you should be aware of them.

Cotton is likely to remain in use for the foreseeable future, though, because as a fibre, it does have some positives. It is versatile, breathable, soft, and clothes made with it do last pretty much forever with care (think of how many ancient t-shirts there are in your closet right now), so the production costs are at least amortized over several years. It is recyclable. And it is possible, with a little work, to source from clothiers that are trying to clean up their chemical processes and reduce water use.

But cotton isn't the only natural fibre, by any means. You can also investigate clothing made from hemp, flax, bamboo, ramie, jute, or abaca as alternatives. These can be more expensive to purchase up front, but we've already discussed how we should be viewing clothing as a long-term choice.

Wool is another option, but as it is sourced from animals like sheep, goats, muskox, rabbits, or camelids, it can also be problematic for a few reasons. Wool has its own colonial history (many indigenous peoples were displaced to make way for sheep pasturing), and the number of resources required to feed and water fibre producing animals is large. There's also the issue of animal welfare, as some producers focus on profit over animal well-being.

So, just as with your detergents, watch out for green washing. Just because something is made from one of these fibre sources doesn't mean it's made sustainably. Take a bit of time to investigate the company and their practices, from farm to the factory to your closet.

Keeping Them Clean

Detergents

The detergent you use in your laundry can have an enormous impact on the environment. Choose detergents with biodegradable ingredients that won't pollute local water sources or cause harm to wildlife. Avoid phthalates and chlorine bleach, which can be toxic to aquatic ecosystems. Look for products labeled "green" or maybe "natural," but always check labels carefully before deciding what to buy, to avoid greenwashed claims. (Also remember that "natural" isn't automatically better. Lava is natural. So are very large tigers. You wouldn't want either of these in your laundry room.)

Avoiding synthetic fragrances. Not only are there an increasing number of people allergic to these things (I'm one of them!), but there is also evidence to suggest that they may not be good for us.

Wash clothes less often

We all want our clothes to be clean and fresh, but you might be washing your clothes too often. Unless you work in a manual labour job where you're sweating a lot or exposed to a lot of odours that cling to you, chances are you don't need to throw them in the wash after one trip out of the closet. This is also true of your bath towels. You're theoretically clean when you come out of the shower, so just make sure your towel can air dry easily and only throw it in the wash after several uses. You'll save energy, water, and money on soap. Not to mention time doing laundry.

Use free drying solutions

If you have the time and space, you can hang your clothes up on a line to air dry them or use one of those indoor drying racks. If you're like me and are super pressed for time, or you're like my in-laws and live in a rainy climate, consider investing in a solar panel and battery setup to provide power for this energy intensive appliance. Speaking of which...

Energy ratings for the win

If you're in the market for a washer or dryer (or both) make sure you're selecting for high efficiency ratings. Avoid natural gas-powered appliances, because methane is a far more potent greenhouse gas than carbon dioxide, and the fewer opportunities for methane leaks there are, the better.

Consider cold water washing

Most machines now come with settings for both hot and cold water temperatures when washing clothes, but cold water can do just as good of a job and uses less energy. Make sure your detergent works in cold water. If you're not convinced by cold water, try warm instead of hot. Compromise!

Ditch the fabric softener

This is one of those things a lot of us buy reflexively because it's how our parents did it, but unless you're drying your clothes on the line... you don't need it! Clothes come out of the dryer softened already as the tumble action ensures they don't get stiff. Liquid fabric softener is rarely biodegradable and often laden with fragrances; dryer sheets are the same, plus the sheets must go in the garbage when you're done.

But the static! I hear you cry. “Dryer balls” eliminate most of this problem. Get one or two wool dryer balls and chuck them into every load. They're reusable and they may also reduce drying time, saving you money.

Washing machine lint

We're used to thinking about the dryer lint trap, but washing machines generate lint too... and all those clothing fibres go into the water system. Even if they're natural fibres, that's a *lot* of lint going back out through your sewer. Check to see if your washing machine has a lint catcher, and if it doesn't, consider purchasing one to add to laundry loads to reduce what gets washed out of your machine.

Ironing

I don't know how many people iron their clothes anymore, but if you're one of them, you can cut down on the amount of ironing you need to do by setting a timer on your phone (or elsewhere) and taking clothes out of the dryer when they're still ever so slightly damp. Hang them up immediately and you'll avoid one major source of wrinkles: clothes sitting around in dryers.

Stain remover

If you're like me, this would be impossible to do without. Fortunately, there are now plenty of options on the market for eco-friendly stain removers, many of which can be bought in bulk or refilled for better packaging options. You'll benefit here through lower costs and from fewer synthetic perfumes.

Instant or tankless hot water heater

I covered this in the bathroom and kitchen section, but it applies here too. A "tankless" or instant hot water heater heats what you need when you need it, rather than expending energy all day to keep hot water available. They're more expensive than regular hot water heaters, but they will pay you back in savings.

Garage Part II

We talked about personal transportation in the section *In The Driveway*, so why do I have another section labelled garage?

Let's talk about tools and toys. If you have an apartment or other rental accommodation, this section might not apply to you, but give it a read over anyway, perhaps to talk it over with someone who owns a home.

The average suburbanite has a garage stuffed full of tools for do-it-yourself (DIY) work, lawn maintenance tools, camping equipment, perhaps ATVs or snowmobiles... the list goes on and the contents will vary according to the family hobbies.

Now don't get me wrong: I am in favour of DIY and hobbies. But do we really need to *buy* these items?

Think about this: Purchasing a high-quality tool might seem like a solid investment for a weekend project, but how often are you going to use it? Certain tools will get picked up more frequently than others. A set of screwdrivers will come in handy, but a table saw? Unless you do woodworking as a frequent pastime, it will sit idle 99% of the year.

If you hit the links every weekend, a set of golf clubs might be reasonable; they're also something fairly personalized. But canoes, skis, jet skis, and all kinds of other recreational equipment? How many times a year are you using these?

If the answer is, objectively, not very often, then why not rent instead?

You might balk at this, comparing the rental cost of some items to purchase cost and factoring in the warm glow of "ownership." I get it. We like our stuff! But let's factor in some of the other costs.

Tools need to be cleaned, maintained, and kept in a place where they will not get dusty or damp. Storage is another factor. Tools can take up considerable space, a real issue in smaller urban homes or apartments.

Be honest. Where are your cars parked right now? In the garage? Or out in the driveway because there's not enough space for the cars? And the fact that some of us have so much stuff that we're renting additional storage space should tell you something.

Renting equipment eliminates this concern, allowing you to use the space for other purposes when the tools and toys are not needed. Renting also grants you access to high-quality or specialized tools that might be too costly to buy. It's an economical way to use top-of-the-line or specific tools for unique projects without the hefty investment of purchasing them.

As for toys, there's the cost of schlepping them to your venue, and the time to get them into the trailer or the top of the vehicle, take them off, put them on again, and then take them off again when back home. Perhaps more of that time could be used for enjoying the activity?

More than anything else, however, let's consider the environmental perspective, as that's what this book is about. Tool and toy manufacturing and disposal have a significant impact. By choosing to rent, you contribute to reducing this footprint. Shared use of tools means less demand for new products and less waste, aligning with sustainable living practices.

If none of this resonates, then consider a different approach: Saving up and buying very high-quality versions of the tools and toys you want, and keeping very good care of them, so that they can at least be passed down to the next generation. I've had to look after three estates now, and the sheer volume of cheap or poorly maintained material in each of these was heartbreaking. I did my best to disassemble and recycle where I could, and donated dead tools and appliances to repair shops, but just as in our section on clothes, these places are getting overwhelmed. As the Boomer generation downsizes, it's only going to get worse. There's a tidal wave of "stuff" coming.

On that note, another approach is to buy second hand. That tsunami I just referenced is going to make for a lot of bargains, so check out estate sales and yard sales rather than buying new. You might get lucky and get some higher quality tools to boot!

One other thing: do try to be objective about the environmental cost of your toys. Those jet skis, for example, are loads of fun, but they're another source of emissions (unless electric), and direct pollution in the water. Because they're so fast and manoeuvrable, they stir up the water like crazy, and sometimes riders actually strike and injure or kill wildlife. As always, multiply the effects of one person doing something by the number of people doing it, and you can see where certain hobbies are a net negative for the environment. Let me put it this way: how much would you enjoy it if motorcyclists suddenly went charging around your house? Probably not much.

I know, I know. This is why environmentalists come across as killjoys. Please remember that the goal here is not to ruin your day, but to ensure we have lovely wild spaces and waterways for future generations to enjoy. There are lots of ways to enjoy being outdoors and even on the water that aren't as destructive; there are also ways to improve other hobbies, like golf, that reduce their environmental impact. It's just going to take honest conversations and a bit of imagination.

Household Electronics

Electronic devices are ubiquitous. It's not unusual for a single person to have a laptop, a tablet, and a phone, as well as a television. With respect to the phone, it is hard to live in modern society without one.

However, the environmental impact of these devices extends far beyond their useful lives in our homes. Understanding the life cycle of electronic devices is crucial for homeowners aiming to adopt more environmentally friendly practices.

The journey of an electronic device begins long before it reaches our hands, starting with the extraction of raw materials. Precious metals like gold, copper, and rare earth elements are integral to manufacturing electronics. However, their extraction often involves significant environmental costs, including (if not sufficiently regulated *and* policed) habitat destruction, soil erosion, and pollution of water sources. The energy-intensive processes of refining and transporting these materials can contribute substantially to carbon emissions and will do so until we electrify the entire supply chain.

Manufacturing the electronic devices is a complex process that involves assembling many components, often sourced from different parts of the world. This stage of the life cycle is energy intensive and contributes to greenhouse gas emissions. The manufacturing process can cause the release of toxic substances, such as lead and mercury, which can harm ecosystems and human health if not managed.

Once manufactured, electronic devices are packaged and distributed globally, a process that further adds to their carbon footprint, and again, will continue to do until we electrify transportation. During their usage phase, devices consume electricity, contributing to their overall environmental impact if the local grid isn't converted to renewables. While newer models can sometimes be more energy-efficient, the rapid pace of technological advancements (and marketing hype!) means devices become (supposedly) obsolete quickly, leading to increased waste. And sometimes they're not more energy efficient at all, they use more net energy.

Finally, disposal of electronic devices poses significant environmental challenges. 'E-waste,' as it's called, can lead to the release of hazardous substances into the environment.

Let's Fix It

Fortunately, there are already several ways to mitigate the environmental impact of your electronic devices because the materials inside those gadgets are valuable. But let's start with the basics.

As always, when you buy, consider second hand, refurbished, open box etc., first. If there's a choice, look for certifications like ENERGY STAR or EPEAT. Also, if there's a choice, try to source from a company that at least attempts to have sustainable supply chains (and ethical ones because a lot of the countries where materials come from have weak environmental and labour regulations.) I'm keenly aware as I write this that there aren't that many manufacturer or software choices, so it's a case of picking the least terrible option.

Try not to succumb to marketing hype. You don't *need* a new phone every year, and there's no good reason to upgrade from a 55-inch screen to something that takes up an entire wall of your living room. (Even the 55-inch screen seems massive to me, as I grew up with a 21-inch console set!) As for resolution, you can already see every pore on the actor's chin at 4k, so upgrading beyond that seems redundant.

Work on extending the life of your device through basic care and maintenance actions like cleaning, dusting, updating software, tidying up old programs and running virus scans, and using protective cases where appropriate. There are eco-friendly options for those cases that will last longer and be easier to clean than the cheap vinyl jobs you see everywhere.

If your device has issues, please do repair rather than replace. (I realize this shouldn't have to be said, but I know more than a few people who junk at the slightest issue. The expense boggles the mind.) It's usually cheaper to repair, and though it's inconvenient to be without a device for a while, you'll manage! And if there's a genuine reason you must have a particular device at all times — emergency contact type reasons, for example — then save one of your old devices and swap the SIM card into it temporarily.

Unfortunately, many devices are no longer supported by software updates and do eventually cease being useful. When you've wrung as much out of your device as possible, please make sure it gets to an e-waste recycling facility. There are several places in a city to drop these off, so it won't be too bothersome. Don't forget to wipe/factory reset the device and clear any media off it.

Decor, Furniture & Renos

Home Sweet Home

Whether you rent or own your residence, I think it's safe to say that most of us want to live in a space that is clean, comfortable, and at least a little stylish. That's fair and reasonable. Your home is, well, your home with all the personal, cultural, and social importance that the term implies.

Unfortunately, just as with clothing, we've normalized a kind of "fast house fashion." We're encouraged to update our paint colours, or do whole room makeovers, or choose appliances based on a 'look' not just once we move into new digs, but every few years, or even seasonally!

I've even seen a store with a name that encapsulates that whole late-stage capitalism and uber consumption vibe. It's called Decor Envy.

House renovations are particularly bad for waste generation. Contractors rarely take the time to remove things carefully, so things are ripped, broken, smashed, and made unusable. It's all tossed in the same bin and thrown in the dump. The U.S. Environmental Protection Agency, in 2018, estimated that roughly 15 million tons of drywall and plaster gets junked every year. Add another 40 million tons of wood products, 15 million tons of asphalt shingles, 12 million tons of brick and clay tile, and 4 million tons of steel. This doesn't include waste generated during construction (think: carpet remnants or trimming wood beams, etc.), and I haven't touched upon the resources required to make and ship these materials either. This figures apply only to the US, by the way.

Yikes.

None of this waste is particularly pleasant stuff. Everything from the glue in MDF "wood" to the compounds in paint, to the stuff we must apply to furniture to make them less likely to burn... all of that leaches from the debris into the soil and groundwater, eventually.

And let's not forget about the expense! Renovations are not cheap, they rarely go smoothly or stay on budget, and we're encouraged to finance them, which means we're making banks richer.

So, let's talk about how we can ease up on the planet and our wallets.

Renovate Sparingly

It's fine to want nice things. But consider, before you plunk down your hard-earned cash, whether you want those nice things, or whether you've been told — via marketing — that you want *those particular things*. And then consider whether they'll work for you and your situation.

For example, as I write this, the trend in bathrooms is to have shower stalls with marble-look tiles and floor to ceiling glass walls. Sometimes, depending on layout, they don't have a door, they're just walk-in.

And they look amazing! Very luxurious. But when you use one? It's a lot like being in a fishbowl; you'd better hope no one walks in on you. The tiles can be slippery, so you need to put down a safety mat, which kind of ruins the look. Keeping all that glass clean is a chore, especially if you have so-called 'hard' water that leaves mineral streaks. And the walk-in ones? Brr! Very drafty. One I used at a hotel also splashed water *everywhere*, no matter how I pointed the shower head, or even if I reduced the water flow. They do have the advantage of being able to step into easily, which is great if you have mobility issues, but since many don't come with safety rails or benches, still not quite functional in those circumstances.

Especially egregious are those kitchen renos for the latest look in homes where the owners never cook! It's a lot of expense just for show and doesn't do the next owner any favours if nothing is set up to prep and cook efficiently. So have a hard think about *why* you want to renovate a particular space and spend a lot of time thinking about how you are going to use as well as maintain that space. Form *and* function here.

If you do really need or want to renovate, consider the logistics carefully before committing. Do you truly need to gut the space, or can you make use of the existing arrangement? Gutting almost always reveals nasty surprises that you won't have budgeted for, but more to the point, you might save a lot of money and cut waste by refitting what you have. In our current house, the kitchen cupboards were sickly lime green melamine jobs (which right there is a good reason to avoid trends!) and the finishes were cracking and peeling. Rather than rip out the cupboards, because they were still quite usable and well-arranged, we refinished them with a classic oak look that goes with everything and put on new doors. The old doors got reused in a variety of ways.

Furnishings

I'm going to set the stage for this section by first digressing into the topic of boots. Bear with me.

In Sir Terry Pratchett's book *Men at Arms,* there's a character called Sam Vimes. He's a cynical working-class cop (or, because it's British, copper) who has spent a lot of time walking his beat and pondering life's injustices. At one point, he comes up with this gem:

> The reason that the rich were so rich, Vimes reasoned, was because they managed to spend less money. Take boots, for example. ... A man who could afford fifty dollars had a pair of boots that'd still be keeping his feet dry in ten years' time, while a poor man who could only afford cheap boots would have spent a hundred dollars on boots in the same time and *would still have wet feet*.

This has become known as Sam Vimes theory of socioeconomic unfairness, or just "boots theory."

The concept can apply to furniture. Setting up a home is expensive. We usually do it when we're young and not making very much money. And we're actively encouraged to buy cheap stuff that isn't very durable and that breaks in ways that aren't repairable (think: soft pine wood, chipped veneers, particle board or pressboard and other 'engineered' products). When they look too crappy, we're encouraged to bin them and buy again. And so, throughout our lives we spend 10x what we would have on more durable pieces.

The cure? Start by plugging your ears against the barrage of marketing messages designed to make you want to buy new and trendy. All being 'trendy' will do for you is make you dissatisfied with your purchases in a few years' time when they look 'dated' versus all the latest trendy items.

Whenever possible, buy used. It takes a bit more time to find the right pieces to achieve a coherent look, true, but the time you invest now will save you time and money later.

When you can afford it, buy durable. This might mean saving up and being patient or doing without a piece for a while.

If you must buy new, try to encourage better practices in the furniture industry by voting with that wallet. Choose FSC-certified (Forest Stewardship Council), products, reclaimed wood, or consider acacia or bamboo products if made sustainably.

Bedding, Curtains, Carpets, Throws, and Other Textiles

Whenever you can, avoid synthetic fibres here just like you would with your clothes. You won't avoid treated material, because there are fire regulations (for good reason), but you can attempt to source the base material more sustainably.

Picking materials that can also be washed at home (in the machine or by hand) is also better from a cleaning chemical usage point of view. Dry cleaning and carpet cleaning use special chemicals that don't always get treated before being dumped back into our waterways.

Décor

Decor is where your personality shines, and you might consider it a challenge, rather than a chore, to do this sustainably and inexpensively. Start with eco-friendly paint with reduced volatile organic compounds (VOCs), as these paints ensure better air quality in your home. Be sure to dispose of any paint cans, eco-friendly or not, via the hazardous material collection stream of your municipality.

With decorative items and knick-knacks, think recycled glass vases, upcycled art, your own crafts if you're talented, and if you're into bitty flames, soy, or beeswax candles rather than scent plugs or sprays. There are all kinds of options if you look beyond the big box outlet stores, and you'll also be able to craft your own unique look, rather than being a mass market conformist. As I mentioned in another section, the Boomer generation will downsize en masse soon, so you'll be able to find a lot of items on the second-hand market.

And don't forget houseplants! Nature's artwork, purifying the air, and eating up CO2 while adding a splash of green. If you have pets, make sure you have nothing that's toxic to them anywhere in the house; veterinarian's offices usually keep lists of stuff that isn't good for them. Even the pots can be sustainable, made from recycled materials or repurposed items.

Things to Avoid

In terms of sustainability, there are some materials you want to avoid, unless you know the source of the materials. Things made of tropical exotic woods (e.g., mahogany) often have a history of colonial exploitation associated with them. They're mostly hardwoods and are slow growing; harvesting them usually just means deforestation of old growth. Where there are plantations, make sure they're in the native range of the tree.

The same thing goes for incense, which on the surface seems like it would be sustainable. The problem is, it's not always clear whether the scent provided is natural or synthetic. Burning things produces indoor air pollution, and harvesting the wood is driving deforestation and, with some species of sandalwood, threatening extinction. Once again, the law of big numbers: when we're all consuming something, depletion happens fast.

Plastic anything, décor wise, should be avoided. Sure, it's cheap, but it will fade, get brittle and crumble or break, and then it's off to the landfill.

And let's avoid mass-produced items. They also tend to be cheap, made of non-sustainable materials, and... well, they're the same as everyone else has. Shop instead for handmade crafts, artwork, and other items.

Holidays and Life Events

The festive spirit often brings joy and a sense of togetherness, but it can also bring a massive increase in waste and environmental impact. We don't like to talk about this much because who wants to be a downer during the holidays? But talk about it we must. This chapter delves into practical and creative ways to make your holidays and major life events more sustainable, blending the joys of celebration with the responsibility of environmental stewardship.

Rethinking Decorations

Decorations are a hallmark of a lot of holiday festivities, but they often involve materials that are not environmentally friendly. Instead of buying new decorations each year, consider making your own decorations using recycled materials. Old magazines, fabric scraps, and natural elements can be transformed into beautiful, eco-friendly decor. As a child, we celebrated

Christmas, and I enjoyed making paper chains out of red and green construction paper, and popcorn and cranberry strings. With my own children now, we use brown parcel paper and different coloured ink stamps to make our own wrapping paper (which can then be recycled, as the stuff you get at the store usually can't be). Whatever your holiday traditions, see if you can incorporate something a little more crafty than commercial; the added benefit is the family time this involves.

If you also celebrate Christmas, you can solve the debate between artificial tree vs. real tree by getting... a smaller potted tree, which then can be replanted somewhere.

Renting, borrowing, and thrifting are also options. Some communities offer decoration libraries where you can borrow or rent holiday decorations, reducing the need to purchase new items each year. And thrift stores, estate sales, and yard sales often have beautiful decorations for very little money.

If your holiday traditions involve pretty lights, opt for LED lights. They use less energy and last longer than traditional bulbs, and because they don't run as hot, they're a little less of a fire hazard.

Getting Married

Weddings are often decoration intensive. Indeed, weddings are often *everything* intensive, because we've built up a massive industry around supplying them, and we all seem to feel they must be a royal family-level spectacle. Special outfits that we use once, wedding favours, balloons, special invitations, place cards, massive amounts of flowers, ribbons, wedding registries, and let's not forget diamonds and all the ethical and environmental issues associated with this gemstone. I would argue that we spend far too much time worrying about weddings at the expense of the actual marriage, but that's a whole other book. The way we do weddings right now is not very sustainable.

But they could be. I was at a literary-themed wedding recently that used old hardcover books for décor; the books then became wedding favours, so I came away with a vintage copy of a classic novel. I found a lovely dress for my wedding at a thrift shop; my husband rented his outfit. Flowers are at least biodegradable; you could also do potted plants native to your area that people could add to their gardens. You get the idea. It can be as special as you like without being so wasteful.

From Matched to Dispatched

Funerals are a tough subject, and certainly if you've just lost a loved one, the last thing on your mind is environmental sustainability. It's hard enough just going through the motions and processing your grief. The ritual of burying the dead is also deeply ingrained in our religious and cultural history, and very few of us question any of it.

But this is another topic we need to talk about in terms of its impact on those of us still here and those still to come. Funeral caskets use up a lot of resources and they aren't all that great for putting into the ground with all their fancy finishes and synthetics. Embalming uses some pretty toxic chemicals, and burial plots, usually set in manicured grass fields, take a lot of water and mowing. Headstones require a lot of resources to create and maintain. Cremation is an option, but it is energy intensive, and the remains are sterile, so nothing much gets returned to the earth if you scatter the ashes.

In several jurisdictions, regulations are being changed to allow people to choose other options, like being buried underneath a young tree. If more sustainable choices are something you'd consider, then check what's available and make sure you leave instructions for your family in your will.

On to happier topics...

Conscious Gift-Giving

Gift-giving is a tradition in many cultures, but it often leads to excessive consumption and waste. When you have advertisements suggesting that you get your Mum a new car for a present, it's gotten silly.

Instead of physical items, give experiences. Concert tickets, cooking classes, or a day at a local park can create memories without the environmental cost of manufactured gifts (and the storage considerations afterwards for all our 'stuff').

Handcrafted items not only reduce environmental impact but also add a personal touch. Knitted scarves, baked goods, or homemade candles are thoughtful and sustainable.

I talked about how we make our own wrapping paper above, but you can also use reusable gift bags and tissue paper. Bonus: "wrapping" with these is fast and easy!

Greeting cards might not seem like a big deal, but so many of them now come with gimmicks (like googly eyes or musical computer chips) that they aren't recyclable. So, if you must get a real card (versus an electronic one), choose wisely.

Participate in local community events that focus on sharing and togetherness rather than buying and consuming. Indeed, community, volunteering, looking after the poor, lonely, and needy are the original basis for most of our big holidays.

Finally, shop local whenever you can. Those mom-and-pop stores will be very grateful for your support. When you do, skip the plastic or paper bags, and use reusable totes to bring your goodies home.

Loot Bags

It's funny how we've normalized these at kid birthday parties when the very name suggests pillage and plunder.

School and private lesson "treasure boxes" also echo this sentiment.

I get the rationale behind both. Loot bags make everyone feel like they got a gift, and treasure boxes are wonderful motivators for behaviour and accomplishments when you're younger.

But the cheap plastic goodies from the dollar store won't last and become landfill within weeks of you sending them home. And the message implicit is that disposable culture is the norm, and quick dopamine hits are okay.

Pick stuff that's going to provide lasting value, and that is at least recyclable. At birthday parties, maybe some food to send home, or something like a colouring book. For treasure boxes, since budgeting is a consideration, why not have kids collect tokens to exchange for something a bit bigger and nicer?

Sustainable Feasting

Our holiday feasts are historical leftovers, really. Festivals of rare abundance, in times where we didn't have a lot of reliable food and we also didn't have a surfeit of distractions.

Today, most of us in developed countries have access to (relatively) cheap and abundant food most of the time, so holiday meals are often times of excess piled on top of daily overconsumption.

To counter that, instead of making these meals about quantity, make them about quality. Go for super fresh ingredients, take the time to handcraft something beautiful, or focus on flavours over copious quantities of sugar. This would be a good time to try a new cuisine.

Choose local and seasonal produce for your holiday meals. This supports local farmers and reduces the carbon footprint associated with transporting food long distances. And consider incorporating more plant-based dishes into your holiday menu. Plant-based foods have a lower environmental impact than meat and dairy.

Halloween, Easter, Valentine's, and the Rest

I've listed a lot of North American days here because these are the ones I'm most familiar with, and because their central focus on consumption is spilling over into other traditions. The overwhelming message of the marketing around these special days is Buy! More! Stuff! Especially cheap chocolate and candy, which encourages bingeing (which has all kinds of consequences) and produces tons of waste as non-recyclable plastic wrappers.

It's hard to resist the pressure, though, especially if you have little children. Who wants to be the kid of the hippy mom who's opted out?

Here too, the answer is to focus on the experience and quality over quantity. Take the kids out for a short Halloween walk to collect a little candy; make the focus on creating a cool costume that can be reused later. Egg hunts can be about finding egg cut-outs to exchange for one larger chocolate goodie. And so on.

Let's Talk About Fireworks

I like a good kaboom as much as the next person. My dogs? Not so much. I'm betting the local wildlife doesn't like it much either. Between noise, scary flashes, chemical residue and spent packages, fireworks are not a good time for the environment or the wildlife.

Rather than launching your own and annoying the neighbours, go to a professional show. The display will be more impressive, it will be further away from houses and wildlife, and they clean up after themselves.

Drone shows are also now a thing. You can go see drones with bright lights flying in impressive formations without the waste and noise. You can preview what this sort of thing looks like on YouTube (although the videos don't do them justice!)

Where To Go From Here

This book was meant to be a relatively quick read and an answer to the question, "Yes, but what can *I* do?" and especially, "what can I do, *right now*?"

I hope that's what you found here.

My goal was to help you feel and be less helpless, more able to take action. Less dependent on politicians who seem to spend a lot of time jetting to COP, and very little time legislating solutions into existence.

I've only just scratched the surface of what's possible, of course. I've put together a list of books to take you deeper into the topic, if that's where you want to go. Because websites are easier to update than book manuscripts after publication, I've put it on the website for this book. You'll find the link below.

If you did enjoy the book, I hope you'll consider:

- Recommending this book on social media, or personally to friends and family. Hashtag: #letsfixthis
- Leaving a review online if you bought it at an online retailer, or on Goodreads, or your favourite social book site. Reviews help other readers make decisions about what to read!
- Bulk orders. Special discounts are available on quantity purchases by corporations, associations, and others. Contact customerservice@tigermaplepress.com for details.

http://www.letsfixthis.eco

Notes

I. At The Systems Level

1. https://www.theguardian.com/sustainable-business/2017/jul/10/100-fossil-fuel-companies-investors-responsible-71-global-emissions-cdp-study-climate-change
2. https://www.coca-cola.co.uk/our-business/faqs/how-many-cans-of-coca-cola-are-sold-worldwide-in-a-day
3. https://vergenoegd.co.za/sustainability/
4. https://stateofgreen.com/en/solutions/the-danish-deposit-return-system-for-recycling-drink-cans-and-bottles/
5. https://www.yellowstonepark.com/things-to-do/wildlife/wolf-reintroduction-changes-ecosystem/

Local Government

1. https://www.theguardian.com/environment/2023/jun/30/low-emission-zones-lez-improving-health-studies

Regional Government

1. https://www.ncei.noaa.gov/access/billions/time-series

National Government

1. https://ourworldindata.org/grapher/co-emissions-per-capita?tab=chart&country=CHN~CAN~IND

On the Job

1. https://www.ndtv.com/world-news/video-fire-rages-in-the-middle-of-ocean-near-mexico-2478211
2. https://drawdown.org/
3. https://www.bbc.co.uk/news/world-us-canada-57678054

4. https://www.ctvnews.ca/world/plague-of-ravenous-destructive-mice-tormenting-australians-1.5446320

Donations and Investments

1. https://www.charitywatch.org/
2. https://www.charityintelligence.ca/index.php

Listening to Indigenous Voices

1. https://www.indigenouspeoples-sdg.org/index.php/english/
2. https://www.ienearth.org/

Right to Repair

1. https://www.repair.org/

Public Transit

1. https://transalt.org/
2. https://www.transportenvironment.org/
3. https://www.ptua.org.au/
4. https://transitcenter.org/
5. https://www.transportaction.ca/
6. https://www.itdp.org/
7. https://www.sustrans.org.uk/
8. https://enotrans.org/

Labour Action

1. http://www.labor4sustainability.org/

Population Growth

1. https://ourworldindata.org/grapher/co-emissions-per-capita?tab=chart
2. https://www.youtube.com/watch?v=jx85qK1ztAc

3. https://copenhagenconsensus.com/post-2015-consensus/health-women-children

Developing Countries

1. https://www.bcorporation.net/en-us/certification/
2. https://commitwithnphnicaragua.simplesite.com/active-projects/eco-stoves

Social Media

1. https://www.france24.com/en/europe/20220123-we-re-at-war-the-lithuanian-elves-who-take-on-russian-trolls-online

II. At The Individual Level

1. https://www.sciencealert.com/microplastics-found-in-every-human-placenta-tested-study-finds
2. https://www.npr.org/2020/09/11/897692090/how-big-oil-misled-the-public-into-believing-plastic-would-be-recycled

In the Driveway, Garage, or Parking Lot

1. https://www.plugshare.com/
2. https://www.tesla.com/findus
3. https://en.wikipedia.org/wiki/Electric_car_use_by_country
4. https://www.energy.gov/energysaver/fuel-economy-cold-weather
5. https://www.theverge.com/2021/7/21/22585682/electric-vehicles-greenhouse-gas-emissions-lifecycle-assessment
6. https://www.nbcchicago.com/consumer/certain-jeep-cherokees-recalled-due-to-chance-of-fire-owners-told-to-park-vehicles-outside/3142517/
7. https://calgaryherald.com/news/world/hyundai-and-kia-recall-nearly-92000-vehicles-and-tell-owners-to-park-them-outside-due-to-fire-risk/wcm/7153f028-decd-4198-86fd-acf5f5a9a0e7?utm_medium=Social&utm_source=Facebook&fbclid=IwAR2h5pzOJEsUAXeuDdFB7cn4hJKIgu15Hb3EJy-E8gXGtNaHOAIpU18AvFiE

8. https://www.teslarati.com/tesla-supercharger-canada-coast-to-coast-timelapse-video/amp/

Garden

1. https://www.wildlifetrusts.org/news/environmental-groups-and-monty-don-write-government-calling-end-use-peat-compost
2. https://www.ecosia.org/

Compost

1. https://www.gardeningknowhow.com/composting/ingredients/what-can-you-compost.htm

Pets

1. https://www.statista.com/chart/15195/wind-turbines-are-not-killing-fields-for-birds/

Power

1. https://youtu.be/SD9yVca6hHI

Travel and Commuting

1. https://www.colorado.edu/mechanical/2016/07/25/how-much-fuel-does-cruise-ship-use
2. https://www.sciencedaily.com/releases/2021/09/210928193815.htm

Kitchen

1. https://www.scientificamerican.com/article/are-gas-stoves-bad-for-our-health/

Food

1. https://ourworldindata.org/less-meat-or-sustainable-meat

Heating & Cooling

1. https://theconversation.com/cooling-conundrum-hfcs-were-the-safer-replacement-for-another-damaging-chemical-in-refrigerators-and-air-conditioners-with-a-treaty-now-phasing-them-out-whats-next-191172

Closets and Laundry Rooms

1. https://theoutline.com/post/2602/clothing-companies-are-trashing-unsold-merchandise-instead-of-donating-it
2. https://www.wired.com/story/fashion-disposal-environment/

About the Author

Chandra Clarke wears many hats, sometimes all at once, which makes it hard to get through doorways.

A long time environmental advocate, she currently volunteers with the David Suzuki Foundation in the Butterflyway project, and is a member of Rotary International, serving as environment chair in her local club.

She is also recovering/relapsing entrepreneur, and a freelance writer. In addition to her nonfiction work, she writes hard science fiction, and humorous fantasy.

Her main website is www.ChandraKClarke.com.

Made in United States
Troutdale, OR
11/30/2024